Purr-fect Voices

A DEEPER UNDERSTANDING OF ANIMALS
& TELEPATHIC COMMUNICATION

 *Jenny* SHONE

This is a work of nonfiction. The information in this book is based on the author's knowledge, experience and opinions and is true and complete to the best of the author's knowledge. Whilst every care has been taken to check the accuracy of the information in this book, the author disclaims any liability in connection with the use of this information.

First published by Jenny Shone, 2022
PO Box 464, Walkerville, 1876, South Africa
E-mail: jenny@animalhealing.co.za
www.animalhealing.co.za

Copyright © 2022 by Jenny Shone

Editor: Jennifer Mathews
(www.ballpointpencil.com)

Cover Design and Interior Formatting by Gregg Davies Media (Pty)Ltd
(www.greggdavies.com)

All rights reserved.
The moral right of the author has been asserted.
No part of this publication may be reproduced, distributed, or transmitted in any form or by any means, including photocopying, recording, or other electronic or mechanical methods, without the prior written permission of
the author, except in the case of brief quotations embodied in critical reviews and certain other non-commercial uses permitted by copyright law.

ISBN: 978-0-620-99191-9 (Paperback)
ISBN: 978-0-620-99192-6 (eBook)

Additional copies of this book can be purchased
from all leading book retailers worldwide.

Praise for Purr-fect Voices

"If you have one or more beloved animal companions, you'll gain new insight into their world and how you can nourish your relationship with them. You'll gain a deeper understanding of all animals and ways of communicating with them. Jenny opens a window into the nature of different species, and the stories are both informative and delightful.

Learning how to work telepathically with all species of animals, explained step by step, will have advantages you may never have considered. To me, it's clear that embarking on this adventure strengthens one's connection not only to the animals in question, but to oneself, to others, and to our world."

– **Mariette Snyman,** Journalist and Podcast Creator

"Jenny helps us understand how communicating with animals improves our relationship with them and enhances our respect for them as spiritual beings. She reiterates how we all have the ability to develop our telepathic skills and includes simple exercises on meditation, intuition, clairvoyance, clairaudience, and clairsentience.

In this book, she shows us how animals help us balance our energy, assist us with our healing, and how they heighten our capacity to love and experience joy. I found this a truly beautiful and enlightening read."

– **Keli Smith**, Part Owner of Pet Chefs Holistic Dog Food, Healer and Author

"This book is a must-read for anyone who ever loved and lost a pet. You will be thrilled by the stories Jenny captures in her second book. Jenny's delightful sense of humour adds to the pleasure of reading this book. Titles such as "A Very Clever and Bossy Pig" and "The Sheep Joined the Conversation" prove that Jenny enjoys every minute she spends with her clients – and their humans. This book will benefit every person striving for spiritual enlightenment and insight into the wisdom animals hold."

– **Sonette Ellis**, Television Producer

"What a fascinating read. I loved it from start to finish. Jenny's writing is warm and heartfelt. Each story brings us closer to understanding the world of animals and our own lives. The book has inspired me to try harder to communicate with my own dogs and other life forms around me and has pushed me towards being a better person. A must-read for those with sensitivity, intuition and a desire for better communication. I found it hard to put down."

– **Sue Green**, Owner, Handler and Trainer of the JOY Dogs: Jade, Orion, and Yola

Contents

Acknowledgments ix
Author's Note xi
A Special Message from the Author xiii
Introduction xv

1. THE UNDERSTANDING — 1

2. THE FANTASTIC FELINE — 4
   The Domestic Cat — 4
   Priscilla's Wisdom — 8
   Smokey — 9
   Samantha — 11
   A Happy Ending — 13
   Schnoekie's Incredible Lesson — 15
   Schnoekie's Message for Laverne — 18

3. EXTRAORDINARY CANINES — 22
   The Energy of the Glorious Canine — 22
   A Very Special Dog — 29
   Winnie's Message — 32
   The Fighting Dogs — 35
   Oliver — 36
   Biggles and Family — 37
   Riff-Raff and the Visualization Technique — 39
   Sceptical — 40
   Story from Micky — 43

4. SOME EQUINE MAGIC — 46
   Some Horse Messages — 49
   A Horse and His Lady — 52
   A Dissatisfied Pony — 54

Healed by a Donkey	55
Rebecca's Lesson	56
5. ANIMALS AND CRIME	58
Stolen	59
Armed Robbery	61
Housebreaking	62
Murder	63
The Guard Fish	65
6. SOME OTHER ANIMALS SPEAK OUT	66
The Parrot	66
A Very Clever and Bossy Pig	67
The Sheep Joined the Conversation	68
Bird of Prey	69
Gecko	70
The Swallow	71
A Swallow Blessing	73
A Message from the Ocean	74
A Conference with a Mischief of Rats	76
7. THE TRUE CONNECTION	78
8. COMMUNICATING WITH THE LITTLE ONES	82
9. THE HUMAN-ANIMAL BOND	87
Animal's Prayer of Love	89
10. DEVELOPING YOUR INTUITION	91
Stimulating the Pituitary & Pineal Glands	92
Relaxation Meditation	93
11. HELPING MISSING ANIMALS FIND THEIR WAY HOME	96
Three Missing Dachshunds	97
Getting Bella Home	98
A Missing Kitten	99
Benjamin's Tale	99
Stolen Husky	100

The Importance of the White Light	101
Guiding a Young Eagle Home	102
12. THE SPIRITUAL SIDE OF ANIMALS	103
About the Author	107

This book is dedicated to the memory of my mother.

*Thank you for all your love and support throughout my life.
I will always appreciate how you allowed me to be me,
no matter how weird or quirky I was.
You encouraged me to grow and follow my path without judgement.
My deep love and understanding of animals came
from witnessing the same qualities in you.*

I love you, Mom.

Acknowledgments

I am grateful to all the animals who have touched my life, helping, supporting, and inspiring me to write this book.

To all the people and animals whose stories feature in this book – thank you for allowing me into the depths of your lives to be able to share your experiences with the rest of the world.

Author's Note

The telepathic connection people have with their animal companions is significant. Understanding animals on a spiritual level is such a privilege, and it is an ability that every person is born with and can develop to a much deeper level.

My journey with animals has taken me to some truly amazing places to meet some remarkable animals and their human companions. I have had many profound experiences connecting with both domestic and wild animals.

If you aspire to connect with beings other than humans, all you need to do is open your heart and take the first step towards making that happen. Communication with animals is incredibly rewarding, and it requires effort and an open mind.

Understanding animals is not just about knowing them better but also to help and heal them when others cannot hear their voices. This knowledge will change your life and improve the lives of all the animals that cross your path.

My wish is that you will enter this magical world where people can talk to animals and hear them in return.

A Special Message from the Author

Please note that although I have connected with many wild animals both in their natural wild habitat and in captivity, I need to stress that I do not support the breeding of wild animals in captivity. However, wild animals bred or kept in captivity do need our support, respect, understanding, and energy.

When connecting with animals in the wild, I have found that they carry clear, strong, and raw energy and are extremely connected to the universe on a deep and spiritual level.

Animals living in the wild often speak to me about humankind and how they want us to respect each other and all living things, including plants, trees, flowers, other animals, as well as our human counterparts. They regularly speak about developing our spirituality to become more tolerant of others.

The animals in captivity are more susceptible to picking up the energies of the people working with them, sometimes even becoming contaminated by the negativity and stresses they feel from the people around them. They need to feel noticed, appreciated, loved, and above all, respected.

When connecting on a telepathic level with domestic animals, they interpret any fear they feel from us as a danger in the area. They assume that we are feeling anxious because of this danger, which sometimes causes them to close themselves off, making it difficult for us to connect on a telepathic level with them at that time. The fear they pick up coming from us could simply be fear or insecurity we feel, not knowing if they will hear us when we communicate with them, or indeed whether they will actually respond in a way that we can hear them.

A wild animal will interpret our fear as hostility and respond accordingly by becoming aggressive towards us or anyone nearby at the time. This, in turn, makes us more fearful and the animal more agitated. For this reason, when connecting with any animal, whether it is domestic or wild, it is vital to keep as calm as possible so that the animal becomes calm. This will lead to more successful communication and the results far more helpful.

Animals teach us about our purpose in life and show us how to reach our full potential as spiritual beings. We can then live in harmony with all living beings – wild animals, domestic animals, family, friends and all of humankind.

The most important message from all animals is to replace our fear with love and respect and then open our hearts to those around us and become one with nature.

The stories in this book are all true. Only some of the names have been changed to protect the privacy of the people and animals involved.

Introduction

B-B

It was the 20th of December 2005, and I had just published my first book, "Paws & Listen to the Voices of the Animals". Everyone asked, "So, when are you going to write the next book?"

My answer was always, "After I have had a good rest," thinking along the lines of about a year. Paws & Listen had taken two years of

INTRODUCTION

hard work, and I hadn't given much thought to the next book. There was plenty of time, or so I thought.

Christmas 2005 came and went, and I was in my element with Paws & Listen doing extremely well. With a new year beginning, I decided to visit B-B, one of my lion friends I had grown close to over the years. B-B lived in a large enclosure with his pride of about sixteen other lions. However, he was not born in captivity and was still very much a wild lion. Hoping to spend some alone time with him, I drove into this enclosure, just wanting to enjoy his company. He was right at the back under some bushy trees when I found him. Parking near him, I switched off my engine and started to relax.

Suddenly he looked at me, and I heard the words, "I am your teacher; you have much to learn, and I will teach you. It is time to write the next book, and I will be a big part of it."

So much for my year's break. It hadn't even been two weeks, and here I was about to start on my second book. B-B had big plans for me.

Over the next few months, I would visit him, not actively getting information from him but just being in his presence. However, I found I just couldn't get started. Some months passed, and B-B and I developed an exceptional relationship. One Monday morning, at the crack of dawn, I awoke with a start.

B-B had connected to me, and I heard the words, "If you don't start getting information for your book now, it will be too late. I am not going to be around much longer."

I assumed he meant he would be crossing over soon as he was not a young lion. This thought was distressing since I wasn't ready for him to leave. I set off to get information from B-B. We had officially started writing the next book.

These words from this amazing lion rang in my head and through my body like an electric current.

He said:

"In ancient times when lions were free to roam the earth, they spread their energy, and in doing so grounded the energy for all other beings.

The lion listens to his fellow beings and acknowledges their existence. A lion only attacks when threatened or for the necessity of food – never out of aggression. On the other hand, humans attack and destroy what they fear or don't understand, usually fuelled by ego, arrogance, and a desire to control.

The heart of a lion says: Love all. Accept all. Respect all without judgement or ego. Listen and hear what others are saying. Within the words, there is always a deeper meaning

Look beyond and learn from others.

Every human, animal, insect, and plant is connected. As you walk the earth along with all other beings on the planet, you realise that you are linked to the same earth. As you look above and see the birds, you realise you are connected to the wind. As you smell the fragrance of the flowers, plants, and trees, you acknowledge that you are connected to the very same planet you share with the animals.

Have you ever watched a pride of lions? Notice the connectedness and the love and respect for each other. Notice how they observe everything around them. They might look as if they are asleep, but they know exactly what is going on around them at all times.

Humans can learn much from lions. Humans need to take part in life. Live life. Be curious. Explore. Feel. Hear. See. Don't shut yourself off from life in a world of stress and anxiety. Be proud of this world we all share.

If you want to stretch, then stretch to the fullest. If you're going to lie in the sun and sleep, then do that. If you want to play, do just that and have some fun. You will be awake, your soul will be happy, and you will experience what true freedom feels like.

Freedom is what you make it. You can live in a small home with limited space, and you can still be free. Freedom comes from within. It is a state of mind, a state of soul. Animals that live in captivity, while this is

INTRODUCTION

not a perfect situation, still have a sense of freedom. Freedom is something you feel in your soul, not merely a physical state.

In an ideal world, all animals and humans will walk freely together on this earth and feel safe and content. However, this is not a perfect world, so we have to adjust our ways and learn to live separate lives and be happy. The day will come when all humanity will be free, and we will all walk this planet together once again.

The three primary emotions to focus on are: Love. Trust. Respect.

Anger is the most destructive emotion. When we feel anger, we hate and destroy the very things we love. Our planet is too beautiful, too unique, too powerful for us to allow our anger to destroy it.

Love our planet. Trust your feelings. Respect all living things.

If one human hugs a tree, respects an animal or loves another human, this love will spread throughout the planet and all who inhabit it. Everyone will experience the feeling of inner peace.

Would this not be worth it?"

This message from B-B confirms just how connected the animals are to the planet and the people who share it with them. It always amazes me the depth of their connectedness. We should take a few moments every day and "listen" to the voices of the animals around us. If we allow them to share their wisdom with us, we will start to see the changes in our beautiful planet that we genuinely want to see. We will become a significant part of these changes.

The animals around us walk this earth without any ego or hidden agendas. They live in the moment and experience everything in such clarity. They carry with them a "raw" energy that is extremely powerful. This energy only becomes contaminated once they join our lives and take on our anxieties and stress.

Humans are generally not good at listening. Now is the time to start listening to what the animals want to tell us.

CHAPTER 1
The Understanding

The acknowledgement of all animals' essential spiritual nature has formed the core of our ability to understand each other. This is due to their unique ability to mirror our emotions back to us in subtle ways. Yes, animals are quite different from humans with their unique bodies, genetic backgrounds, and senses. They experience the world in their own way, which is often dissimilar to the way humans experience the same world.

For instance, a canine species has specific characteristics and behaviour patterns, some of which are opposite to those of a feline or an equine species. It is important to know that when communicating with a dog, you need to consider the characteristics of the species and remember that they are individuals with their own thoughts, feelings, likes, dislikes, and sense of humour, just like humans. We need to see each animal as an individual and as a whole entity spiritually. Not separating the physical and spiritual will give us that deeper connection and understanding of them.

When you make a spiritual connection with an animal, it is almost as if bodies disappear. You see the animals as vehicles of life in the physical and individual expressions of divine creation. When the essential spiritual bond is made, being to being, recognition of likeness – even oneness – occurs. It is magical to feel this deep affinity and respect. It deepens our attachment and opens our channels of communication, trust and understanding of one another.

We cannot reach this level of communication if our relationship is based on sentimentality, and we treat our animal companions as babies or dependant underlings. Our relationship should be a blend of compassion and kinship. If you have experienced this type of soulful union with another creature of any kind, even human, you acknowledge each other as spiritual beings by your approach or attitude. The animals always know.

It does not mean that all animals will communicate with you or respond to your contact since they have their own minds and can make their own decisions. Animals have conditioned reflexes and fears from centuries of experiences relating to their species and to them personally. Their instinct to flee or attack might be according to their biological chemistry and their function in nature's pattern.

Some animals will recognise you as a spiritual being and be more willing to connect with you. Others just don't want to or don't particularly see the need to relate to another species of any kind, including humans. However, if you remain quiet and unobtrusive, respectful, attentive, and ready to connect, most animals will eventually be interested and accept you as part of their environment.

Fortunately, most domestic animals living with humans all their lives will not see them as predators. Therefore, they will be far more willing, comfortable, and ready to connect to human beings.

Bonding with wild animals is a little more challenging. Wild animals would include birds, rats, snakes, lizards, worms, bees; any creature you might find in your garden or in nature that does not fall under the category of domestic animal. To have a wild animal sit comfortably in the presence of a human and connect is a great privilege not to be taken for granted.

The first step to communicating effectively is to understand the species. Once you learn about the characteristics and behaviours of a specific species, the animal feels that you respect them and helps them form a trust which will lead to a meaningful connection.

In the following few chapters, we will take a closer look at three of our domestic animal species and see just how they have adapted to living in our world and what it is that they are trying to teach us.

Then we will look at some of the wilder animals, insects, and the smaller creatures that we often brush off as nuisances.

> "To understand is to accept, and to accept is the beginning of making the perfect connection, which in turn leads to having the ultimate communication."
>
> — JENNY SHONE

CHAPTER 2
The Fantastic Feline

THE DOMESTIC CAT

Even though scientific evidence suggests we are all born with intuitive abilities, in some, this sense is more developed than in others. Humans tend to have little understanding of psychic phenomena and have little trust or belief in the process.

Thirty & Rikki

On the other hand, cats are extremely psychic, and their sixth sense is naturally more highly developed than ours. Cats tend to trust and react to their intuition immediately without first thinking about it or analysing it as we do.

Cats have much mystery surrounding them and can sometimes be a little aloof. Some people feel that because their cat may not come close to them while they are going through a traumatic time, this indicates that their cat doesn't care. It is quite the opposite. In situations like this, the cat has so much respect for their human that they will often give them the time and space they might need to get over whatever is causing them stress.

However, in cases where the human is ill, a cat will often spend a lot of time with them sharing and balancing the energy of the human so that they can heal themselves. Sometimes the cat stays with their human, comforting them as their time for crossing over approaches.

These cases show the ultimate respect that cats have for human beings.

Cats are highly attuned to the energies of others. When a stray cat comes into your home, they are often drawn by the energy of someone in the house; someone in the family who needs the energy they can bring. The human could be suffering from an illness or severe depression. Once the cat feels this person is healthy and balanced again, energy-wise, the cat will leave and move on to the next home, where someone else could benefit from the energy he carries with him.

When cats go missing, we often think that the cat is lost, but this is not always the case. One of the reasons cats rarely get lost is their ability to read the earth's magnetic energy field. To a cat, the earth's magnetic energy field is like reading a map consisting of energy lines, vibrations, and scents that they can tap into and follow to find their way home. They literally have their own built-in GPS.

A few years ago, I had a case where I was asked to help find a cat who had gone missing a few weeks earlier. On connecting with this cat, she told me that she had temporarily moved in with a little boy down the road who needed her. The little boy had terminal cancer, and she was bringing him comfort in his last weeks on this physical plane. The little boy, who was only nine years old, loved her and was very happy to have her with him at this difficult time. She also told me that she would go home again once the little boy had crossed over into the light.

Three weeks later, she arrived home. On further investigation, we found that a little boy living on the next road and a few blocks away had passed away. The family reported that a cat had moved in and given him much love, staying with him right up to the end. The cat had made the boy's crossing so much easier for him, as well as for his family. He had crossed over happy and fearless.

Cats also have the amazing ability to recharge energy that has slowed down and become stagnant. They are symbols of adaptability and teach us how to be more flexible in our bodies and our thoughts. One of the messages they often share is that if our thoughts become inflexible and rigid, our bodies will follow, and we will start to suffer from diseases such as arthritis.

Cats are natural-born healers. Apart from the energy they radiate, they heal through vibration when they purr. Allow the vibration to travel through your body, and it will cleanse all cells of toxins, including those related to certain cancers and arthritis or any of the other debilitating diseases we might develop over the years.

It is truly an honour when a cat chooses to live its life with you!

Even though cats have extraordinary healing abilities, they often pick up the issues and fears of the humans they live with.

The domestic cat teaches us a lot about being aware of the environment we live in. Cats are incredibly in tune with their surroundings, and their senses are always alert. Observe them and notice that when a cat sleeps, it sleeps soundly but is still aware of everything going on around it, just like its big brothers and sisters, the lions. When the domestic cat is awake and roaming around or playing, it is always on high alert in case of danger. It is doubtful and unusual for anyone to sneak up on a cat. They will always be aware of you long before you get to them.

Until you make a deep connection with an animal, you never know precisely how wise they are. Cats are great teachers. Sometimes, through the most painful and difficult situations, they teach the biggest lessons. In most cases, the human will only learn these lessons through dramatic circumstances.

As you will notice in some of these stories, when an animal, or a human for that matter, crosses over, the soul never dies; only the physical body dies. Once the soul is released from the physical body, it is still very much alive.

To see a soul would be like looking at a brilliant spark of light, and it is the purest form of energy that you could ever find. In many of my connections with a crossed-over animal, I have been fortunate to see them in their soul form. Most often, animals will appear to us just as we know and remember them in their physical body. However, this is not their actual physical body. It is just a way for us to identify them to know we are connected to the correct animal.

No one can ever tell you exactly what it is like on the other side or what a soul actually looks like. The impressions I get when I connect with the soul of a crossed-over animal is only my perception of it. I don't believe that anything we might experience or see is anywhere close to what it is actually like. We can only see what our human minds can relate to, and I believe the experience is far more powerful and amazing than anyone could ever imagine!

There is no scientific 'proof' that a soul exists or even that life exists after this life. The proof lies in the information we get from a departed animal or human. They often give specific information of certain events and descriptions that only their human will recognise and understand. Some of these descriptions could be about a new addition to the family, or it could be about a move to a new home or job. Often I have been given a detailed description of the garden and the animal's favourite place to play. On one occasion, the animal showed me the flowers planted on the grave in honour of her passing. After reading the following stories, I will leave you to make up your own mind whether the souls of the departed live on.

When connecting with a crossed-over animal, the information can come through in many different ways, in the form of symbols, colours, pictures and even feelings that go directly to your heart. It is then up to the person connecting with the animal to use their intuition and interpret the messages shared by the animal.

PRISCILLA'S WISDOM

In this story, we see just how profound the soul of a cat can be. Priscilla was a fantastic cat while still in her physical form and living in the physical world. Now that she is in her soul form and living in the spirit world, she is still an amazing cat with such wisdom.

Priscilla

The day Priscilla crossed over was extremely devastating to her human companions. After she had been gone a few weeks, the family asked me to connect with Priscilla and see if she had a message for them. The family missed her terribly and just wanted to know that she was fine and happy.

It didn't take me long to connect with her. She was relaxed and ready to chat.

She showed me water bubbling up from the ground. The water burst into a magnificent sparkling fountain that travelled up to the sky, then came showering down in little sparkling stars that covered the ground.

I asked her what this meant, and she said:

"Think of life as a fountain of water, constantly flowing. When the flow of water stops, it becomes stagnant. The same happens with life. If you stop, you become stagnant and stale. As long as you keep moving, you will keep growing, and you will sparkle with life and energy.

Be the fountain that joins the earth with the sky and comes cascading down as tiny sparkling drops of energy covering the earth and stimulating new growth.

While the body might become weak like the water that lies still, the soul keeps moving and sparkling like the fountain bursting out of the earth. The soul never lies still. It is constant, and it is powerful. It moves through all dimensions of life and the afterlife.

Hear my words.

I know these things."

The family understood that she was well, and they should not become stagnant without her, but rather sparkle in her memory – what an incredible cat with a special message.

SMOKEY

My friend, Alba, had a young cat called Smokey that everyone loved. His passing was very traumatic, and one day while I was visiting her, she asked me if I could connect to Smokey and see how he was.

A few days later, I sat down to connect to Smokey, and he came through almost immediately. He looked regal and said he was an old soul, which means that he had had many lives on earth and was quite advanced spiritually. He had powerful energy around him, with many cats and dogs surrounding him. He said he was in charge of them and was their superior.

Smokey had a special friend with him, a shaggy black dog. I later discovered that this dog had crossed over quite a few years before Smokey, and they had now met up again on the other side.

While talking with Smokey, I kept feeling flashes of female energy, which usually indicates sensitivity. He told me he was functioning as a rescue animal, and his job was to help other animals' souls cross over smoothly when their time came. He said they needed him, which is why he felt he needed to leave the physical dimension early.

When I asked him how he had crossed over, he replied, "How I went is not important."

Once an animal has crossed, they rarely show any concern for how they went. The only important thing is that they are now on the other side. However, while talking with Smokey, I got a tight feeling in my throat.

Alba later confirmed that Smokey had crossed due to a collapsed trachea.

Smokey shared a special message for Alba:

"I love my mom, and I'm proud of her. I feel so light, so happy. I love it here.

Beauty is only skin deep; it's what's on the inside that counts. Look deep within yourself, and you will find great wisdom and knowledge. Use it, for you have remarkable abilities that need to be discovered. Trust and go with love, and I will watch over you always."

One year later, I needed more information from Smokes, as he likes to be called. I asked him to tell me more about his job as a rescue worker.

Here is his story:

"It's a gratifying job, and I love it. I help all animals that arrive here.

Some arrive after passing naturally, and I welcome them home, helping them readjust to their lives here. Those that pass after a long illness or a sudden accident, I take to a safe place, almost like a hospital where they get put into a sleep state, and we do healing on them, helping them recover. This is why some animals don't communicate straight after passing. They are in a state of healing.

Others arrive here and get straight into work the way I did. Not all animals know what their jobs are until they get here. Not all animals are rescue animals; some are companion animals, some are teachers. Just like humans on earth, they all have different jobs. There are many jobs for the animals here. At the moment, I am working with humans and animal babies. I have moved on but only to a different duty, not another place. I still come to visit my earth family at times, and I will have another earth experience but not just yet. At the moment, I'm just too busy and loving it.

Losing an animal friend, although devastating, is part of the human experience. It is essential to your spiritual growth. Even if humans know that their animal is a rescue worker, it is still devastating.

My message to humans is to communicate with the soul of your animal. Feel the love and celebrate their lives on earth and over here. Do not feel anger or pain. Be happy and know that we are happy too."

SAMANTHA

Here is another story where you can clearly see the cat's personality coming through during this reading.

Samantha is a stunning cat, and I felt honoured to connect with her.

When I connect with any animal, I always see what they give me before asking them any questions. This helps build their confidence in me and then be more willing to answer any questions I might have.

Samantha

All animals need to trust the person who is chatting with them. If they don't trust you, then the chances are they will not be open to talking, and you will battle to get anything out of them. After all, why would they share their deepest secrets with a total stranger?

Samantha confidently told me that she is a lovely cat and has a great personality. She could be a little highly strung and nervous at times, but she felt safe living with Dawn and her family. Samantha felt she was an essential part of the family and mentioned that she just loved her name.

Connecting with Samantha, I immediately got the feeling of absolute joy and peace, an overwhelming sense of love. Samantha was quite pushy with me and told me firmly that she did not want me to waffle on. She wanted to get right down to answering the questions, and boy, is she talkative. I decided to leave the letter Dawn asked me to read to her until I had asked all of Dawn's questions first.

Samantha told me that she was sitting on Dawn's lap while she was writing the letter, so she already knew exactly what was in it. She immediately began to speak and told me to relay the message to Dawn and answer her letter.

These are Samantha's words:

"Everybody thinks that when a soul crosses over, it goes to a faraway place in the sky; that is not true. All that happens is we shed our physical form, and our soul goes to another dimension parallel to that of the physical world. That's why it is possible for me to be in this beautiful dimension and also be right there in the earth dimension with you. Humans sometimes call this dimension "heaven", but animals call it "the place of love".

There is so much to do here. I have lots of friends – cats, people and even some dogs. Imagine that! There is no stress, fear, anger, tension, or pain here. There are lots of trees and places to play hide-and-seek. It is such fun. In answer to your question, Yes, Misha and Cleo are here. We all play together.

There is also a large dam with clear water and lots of fish swimming around. I can sit and watch the fish for ages. I never can understand how fish manage to swim. It looks so complicated. Then I suppose they wonder how cats can climb trees and how dogs can bark.

I started a special project when I was still in my physical form. I have created a silver thread of energy that I place right in the hearts of select humans. I then stretch this thread out from their hearts and place it in the hearts of their animal companions. Humans develop a much deeper understanding of their animal friends, and it helps the animals feel their love. Not all people love animals the way they should. I am working on changing this, and it is my way of healing the bond between humans and animals.

Dawn, you need to adopt another kitten. You have so much love to give, and there is a kitten out there that needs it. I know precisely how special I am to you, and this will never change. However, the kitten will be just as special in a different way. I will help guide you to the right kitten, and you will know it when you see it.

I am part of you. Our bond is so strong, and I will always be nearby. As soon as you think of me or call me, I will be there. I will watch over you and the kitten. You looked after me so well while I was still in my physical form. I love you for this. Now it is my turn to look after you.

There is no such thing as "unresolved emotional baggage"; it is all part of the human experience. You learn from every experience in life, some of which are not so good, but they are necessary for your spiritual growth. Without overcoming issues, you will never grow. It is only through "dealing" with the problems that you discover how to move beyond them. Yes, I did feel your emotional state, but now I understand and can help you deal with the emotional issues holding you back.

Of course, I can hear you. And I love it when you sing to me. I can also see everything you do. The vale between our two worlds is very thin. The only reason you can't see me is that the earth's energy is exceptionally dense. If you focus on meditating and concentrate on your breath, you will lighten the energy around you, and it will be easier for you to see and hear me. You are an advanced soul and quite spiritual. The only thing hampering you is when you get caught up in the issues that go with living in the physical earthly realm.

I am one of your guides now and will always look out for you.

We have been together many times before, and we will be together again."

Usually, an animal will give me some thoughts and quite a few pictures or symbols, which I have to interpret. However, Samantha gave everything in her own words.

A HAPPY ENDING

My client living in the UK was in a dilemma. Mandy's beloved cat had crossed over two months before, and she was now ready to adopt another cat from a shelter. Mandy went to the shelter, and while she was walking around looking at all the cats, one little cat stood out for her. She didn't go into the kennel but sat just outside. After a few minutes, the same cat came up to the fence and looked deep into her eyes.

Mandy instantly fell in love with the little cat and felt such a pull from her heart towards this fluffy little being. However, when she enquired in the office about adopting, she was told that the cat named Lightning was not up for adoption. Mandy was devastated, and she asked why. Lightning had come from an awful situation and, as a result, was quite vicious. Apparently, she would often attack the other cats and even the people trying to feed and look after her.

Mandy left feeling saddened and asked me if I would connect with Lightning and find out why she was behaving in this way. She also asked me if I could explain to Lightning that she had to be gentle to get the perfect home and belong to a family that would keep her safe and where she would be happy.

The next day when I connected with Lightning, she felt so nervous. She didn't know what I expected from her, and she was extremely suspicious of me and my motives.

I put her at ease, saying that she didn't need to tell me anything she didn't feel comfortable sharing with me.

Lightning then told me a story of how people had always let her down and mistreated her, showing no love, kindness, or respect. She eventually ended up on the street where she was hungry, cold, and afraid.

Her anger and aggression towards others were brought on purely by her fear and insecurity. I found her to be quite a timid cat and very gentle. She just needed to build her trust in someone and feel their love. I told her about Mandy, and said she wanted to give Lightning a good home where she would feel safe and loved. I told her Mandy would come and visit her the next day and how it would be wonderful if she would allow Mandy into her kennel area and be gentle and loving towards her.

I also suggested that Lightning allow the people working at the shelter to help her by feeding her and being there for her.

The following day Mandy arrived to find Lightning sitting at the fence waiting for her. As she went in, Lightning immediately came to Mandy and rubbed herself against her legs. When she sat in the centre of this kennel area, this scared little cat jumped onto Mandy's lap and purred while she stroked her.

Mandy established that from the moment I had finished chatting with Lightning, the keepers had gone in to feed her, and she was reasonably calm while she waited for them to leave her to eat her meal.

The shelter was so pleased to see the change in this little cat that she was allowed to go home with Mandy a few days later.

She is now comfortably settled into Mandy's home and even sleeps on Mandy's bed with her at night. Lightning has not attacked anybody or any other cat or dog again and is happy being an only cat in a loving home where all the attention is on her.

The moment Mandy and Lightning's eyes met, they had an instant soul connection. They were there for each other at a time when they both needed help. Mandy and Lightning had been feeling lonely, and now they had each other and so much love to share.

SCHNOEKIE'S INCREDIBLE LESSON

Many years ago, a friend of mine, Pat, came to me and mentioned that his friend named Laverne had taken in a little black kitten. The kitten's name was Schnoekie. He told me that Laverne and Schnoekie needed my help. Schnoekie, the Panther Cat as he became known, had a gentle and most endearing nature, and everybody loved him.

Schnoekie, "The Panther Cat"

Then one day, the struggle began. Schnoekie became ill, suffering from bouts of nausea, vomiting, diarrhoea, lack of appetite, his once shiny black coat now lacking lustre and in poor condition. This became incredibly stressful for Laverne, who doted on her little Panther Cat.

For more than a year, a string of vets could not seem to get to the bottom of the problem. There were good days and bad days, various diagnoses and treatments. The consensus was that the little cat's health was compromised, and Laverne should not expect him to live long.

At this time, Pat introduced me to Laverne. After a lengthy discussion, I offered to connect with Schnoekie to see if we could determine why this was happening and if there were anything that this special little cat needed that would help.

The message that came from Schnoekie was not about how we could help him get better; it was all about the gift Laverne had as a healer. He said she needed to follow her passion and learn how to communicate with and heal animals.

Laverne started attending my workshops and taking part in my courses relating to animal telepathic communication and healing. As she grew and developed her abilities, she gained a profound understanding of her powers.

After years of working side by side with me, she started her own business where she would connect with animals on behalf of people to help heal them. This was the incredible gift Schnoekie had given to Laverne, getting her started on an amazing journey of love and healing.

I invited Laverne to share her experience with Schnoekie. This is her story:

"Schnoekie, a miracle cat, lived a whole two years after the diagnosis of severe kidney and heart disease by the specialist vets, radically defying medical science. I am convinced that this would not have been possible without the Radionics, Reiki, love and unique treatment he received.

Then one dreadful day, I came home to find him suffering severe and recurring seizures. I rushed him to the Emergency Veterinary Hospital down the road. After a three-and-a-half-day struggle in intensive care, with confusing periods of despair and hope, sure deterioration and then miraculous rebound, he was finally released to kitty heaven. True to his spirit, Schnoekie, the Panther Cat, did not go without a fight.

I was with him every minute permissible. Hours before his sudden and final exit, he was full of life and energy, convincing everyone that he would be discharged the next day. But this was not to be. His little heart and kidneys could not sustain him any longer. He developed water on the lungs, making his breathing even more laboured. He was barely conscious and did not move at all for many hours.

In desperation, I held the tube providing oxygen to his nostrils in case of another miracle. It was then that I realised the fine thread holding him was diminishing. With the growing distance I felt from him, I knew I faced a heart-wrenching decision. After many hours of silently laboured repose, at the moment when he was ushered from this world to the next, he finally gave up the fight and released his soul into the spirit world.

Schnoekie, the Panther Cat, had just celebrated his sixth birthday, outliving all expectations. 11 April 2011 was the day he transitioned from the physical world. Even though I miss him more than words can express, his soul remains close, and I know that we are separated only by a thin dimensional veil.

This beautiful soul changed my life forever and as a result of the experience, so much more has meaning. It is impossible to ignore that the Panther Cat came into my life for a reason. My instinctive commitment to this phenomenal cat's journey, together with Jenny's guidance, has shown me my passion for helping an animal heal. I learned that healing is not necessarily curing, but sometimes just assisting with the energy to ease the way as far as the physical and soul levels will allow at the time. It is about making a difference, making things a little easier, and providing the love that allows the healing deep within for all.

We all have our path, but we also have the privilege of touching another being's life with permission and allowing each other to grow. This experience has inspired me to learn more from Jenny about the art of animal communication and healing and investigate the fascinating realms of loss and gain, death and life, illness and health.

The experience has been the catalyst for my own profound personal healing journey, which continues to unfold, revealing new insights with time. I am humbled when I comprehend the extent of the gift this special cat gave me. The depth of my healing directly from my experience with Schnoekie can be likened to the development of a chrysalis to a butterfly – challenging and then spectacular. Once the Panther Cat had crossed over, through Jenny, he gave me a profound message with this special symbolism which reinforces the image for me. Cycles that always seem to bring me back to experiences with him.

I always dreamed of helping animals but could not have anticipated how soon and to what extent this would happen. The journey has moved me through spaces I never thought possible. It has sparked new interests, avenues and possibilities, enriching my life in unexpected ways. The inspiration of healing now reaches further to other animals, the entire animal kingdom, humanity, our planet and beyond – a ripple effect with my ever-changing perspective.

Through losing my beloved Panther Cat in this reality and in trying to come to terms with the devastating loss, I gained precious insight from someone who has since taught me much. I came to realise that although in my mind I longed for my precious Panther Cat to be back in my life without any suffering and pain, I could feel and experience him dancing in my heart forever."

SCHNOEKIE'S MESSAGE FOR LAVERNE

Laverne asked me to connect with Schnoekie when he had crossed over.

Every time I connect with the soul of an animal that has crossed over, it makes me realise that this is the reason I am doing this work. I am always amazed at the depth and wisdom of the soul. It is always the most humbling and awe-inspiring experience.

When I connected with Schnoekie, the first impression he gave me was of him surrounded by the most considerable silver-blue light. It was more like glittering sparks joined together, twinkling like tiny diamonds. He was right in the centre of those sparkling diamonds. As he moved, the glittery sparks of light moved with him like a wave of energy.

I asked him about the sparkling silver-blue light. He told me that this is what his soul form looks like, and the only reason he appeared in the centre in his physical form was so that I could identify him. He told me that humans always need a physical form to understand what they are looking at.

The feelings and emotions I got from him were not just pure happiness; he gave me a deep sense of contentment and absolute love. It was something so profound that no words could effectively describe it. I had never felt anything like that before.

He then showed me a picture of himself surrounded by large and most colourful butterflies. He told me that the butterflies represent the joy of life both in the physical and after the physical. They also symbolise change – the change in his life and the change to come in Laverne's life. The colours represent the beauty those changes will bring.

I asked Schnoekie about his job. He said it was to help trackers, both human and animal, to find other people and animals buried in rubble after earthquakes and other disasters around the world. He said he guides the trackers to find these souls, some of whom are rescued, while many need help to cross over. This was when he ushered the souls to safety moments before death occurred. He says it was a privilege to have this job, and he took it seriously.

I asked him if he had a message for Laverne.

This is what Schnoekie said:

"Although it is vital to live your life and experience all the emotions of love, joy, pain, anger, sadness and more, these are just bumps in the road that ultimately lead to considerable growth. I had many bumps in my life too. I learned and grew from all of them. It wasn't easy, but it was my choice.

You and I have learnt a lot from each other, Laverne. We are both powerful healers. You might ask, "How can we be healers when we are not healed or well." The fact that we are not well means that we understand the concept of health and wellness. If we were healthy and well, how would we then develop the understanding that is necessary to help others deal with their health issues?

The most important thing to realise is that to heal the body, you first need to heal the soul. The best way to heal your soul is to follow your passion. In pursuing your passion, you will be freeing your soul. In freeing your soul, you will be showing the respect your soul needs to become healthy and happy. You will then once again feel the joy of life.

Remember, the soul has no fear. It will take on any challenge presented to it. Fear is a physical emotion and can be extremely damaging.

You don't have to accept the bumps that life has put in your way; you can change these experiences. The best way to change the things that hamper your journey is to accept and acknowledge these things and not allow them to become problems. Simply let them go. It is not easy, but you have the power to do this. By allowing bumps to become issues, you will be keeping them with you. By letting them go, you will be clearing them out of your system and your energy field. This will make space for more positive experiences to come into your life.

Laverne, I will always be there to help you. You can't see, feel or hear me but just know that I am with you – I will always be. Your vision is clouded by anxiety and pain, but you will become aware of my presence again when this clears. Our love and commitment to each other are so strong that not even my passing will separate us. Our souls are on a journey together – you in the physical and me outside the physical.

I watch over you every night while you sleep, and this is my choice. Life is full of options, and you need to make the choices that make you happy. Please don't despair; I am near. This is a big bump that you are going through, but it will end, and you will find happiness again. You will allow the smile on your face to come from your heart."

When you connect with an animal who has crossed over, you are connecting with the soul of this animal. The soul never dies; it is the physical body that dies. At the moment of death, the soul is released from the physical body to live once again in another dimension. It is only through our pain and fear that we create blockages that prevent us from seeing, hearing, and feeling our departed animals.

There is always a feeling of excitement, love, and total joy I experience with every animal I have ever connected with that has crossed over. Although it is the most devastating experience to lose a beloved animal, it does bring a little comfort to know that they are still near us and can see everything we do and hear everything we say.

CHAPTER 3
Extraordinary Canines

THE ENERGY OF THE GLORIOUS CANINE

Gabriel

Gabriel came to live with my family and me as a young puppy. I had recently lost my beautiful boy Riff-Raff, who helped me start the Animal Healing Centre. I wasn't even thinking about getting another dog to add to my family at this stage.

One day, I visited my friend who runs the Husky Rescue SA. I suddenly had a strong feeling that there was a little puppy that I needed and who needed me. I asked Joanne if she had any husky puppies ready for adoption. She told me that she didn't keep the puppies at the centre because they were all kept at foster homes until they could be homed. I asked her if she had any males, as I felt so strongly that there was a male puppy out there waiting for me.

It turned out that the foster home had five females and one little boy, but I was informed that the male was already booked to be taken by someone else. I immediately made an appointment to go and meet the puppies.

When I got there, I noticed that the puppies were Husky and Border Collie crosses. They were so sweet. The little boy, who we later named Gabriel, came running up to me and sat right next to me the entire time I was there. I felt devastated that I couldn't adopt Gabriel because I felt such a strong pull towards him.

Later that day, Joanne phoned me to tell me that Gabriel's new home had fallen through because the people had decided to take one of the females instead. I was over the moon with excitement as Gabriel was coming home with me the following week.

Since Gabriel joined our family, he has taught me how to really have fun and even led me to take up dancing with dogs as a fun exercise to do.

I have always trusted my intuition, and if I feel strongly about something, I just know that there is a good reason for it. The energy around us moves in amazing ways and helps us make the right decisions in many situations, as I found out with Gabriel and all the other dogs I have brought into my home.

As you already know, dogs are highly aware and function on a deeply intuitive level. Dogs with whom we share our homes become incredibly connected to us and all our thoughts. We often hear about the ability of our dogs to know precisely when we are about to come home. Many years ago, I put this to the test when I asked Thomas, one of my gardeners, to watch and tell me what time the dogs would go up to the big rock at my main gate to await my arrival.

As soon as I finished with my appointments for the day, I would connect with my dogs and tell them I was on my way home and how long it would take me to get there. About three kilometres away from home, I would connect again and ask them to wait at the large rock as I would be there in five minutes. When I turned the corner approaching my gate, I could see all the dogs sitting on the big rock. Thomas said they went up exactly when I told them I was five minutes away.

I took this experiment one step further and found that on the days I didn't connect with them, I would get home to find all of them asleep in the lounge, surprised to see me. It doesn't mean that you have to connect to your animals to let them know your movements, but the closer your bond with them, the easier it will be for them to naturally feel when you are coming home.

The intuitive abilities of dogs go even further. Some dogs can sniff out cancer, epilepsy, or any other serious condition.

Dogs can also be intensely spiritual. I discovered this on the day Riff-Raff joined my family. The minute I saw him as a spiritual being and not just a beautiful body of a dog, he understood and started growing spiritually at an incredibly fast rate. He would often join me in my meditations and be totally aware of the energy around him and anyone who came close to him. Whenever I took out my crystals, he was there, right in the middle. He didn't want to be left out of anything spiritual that I was doing. All it took for him to realise he had an extraordinary soul was for me to point it out to him and acknowledge him as such. He was by far the most spiritual dog I have ever had the privilege of knowing and working with.

People always say that dogs teach us about the true meaning of unconditional love. While this may be true, it is also not the only thing they are here to teach us – there are many more lessons. There are lessons in loving and trusting yourself, perseverance, patience, acceptance and many more. However, dogs are not only here to teach us but also to learn from us.

Dogs tend to work on a deep level and are great healers, often healing us emotionally. They try their best to make people smile, laugh, and be filled with joy. This makes a lot of sense to me as a healer because most diseases stem from emotional issues. If you can clear the emotions behind the disease, you can prevent it from manifesting into the physical body, and dogs understand this on an intense level.

Baby Riff-Raff as a puppy playing in the mud and connecting with the earth

---

A few years ago, I had a client I was helping with one of her horses. One day, when I finished connecting with her horse, she asked me to help with her German Shepard dog. She told me she worked from home, but every day while working on her computer, her dog would come in, grab hold of her arm, and tug on her until she went outside to play with him.

After connecting with the dog, I discovered that he wasn't demanding a game for himself. He felt that she needed to play because, being so focused on her work, she needed the break before she became ill. This was his way of getting her to leave her computer for a short while and go outside to play and get some fresh air. Yes, our canine companions never fail to amaze me.

A dog's energy field is extremely sensitive, just as ours is. For instance, if someone smokes a cigarette in the vicinity of a dog, or any other animal or human for that matter, it tends to create holes and weaknesses in their aura, which is the energy field surrounding our bodies. This can lead to negativity entering their aura, which can lead to diseases manifesting into their physical body.

Even our emotional issues can affect the dogs closest to us. They don't want us to put them on a pedestal, and they don't want us to look down on them. They want us to walk next to them along our paths so we can have our own human experiences, and they can have their own dog experiences. There is no way that we can have a dog experience, just as there is no way for them to have a human experience. They want to walk side by side as partners and friends.

However, what is important to remember is that we need to stay connected and learn from each other to grow and develop spiritually together. So keep smiling and keep having fun with your doggy companions.

Remember that a dog is a dog; it is not a human. Therefore we mustn't humanise our dogs or expect them to behave or think like humans.

When we bring a new puppy into our homes, it is often the first time they are experiencing all the emotional energy surrounding a human. As dogs are susceptible to energies, this can be overwhelming for them. We then proceed to talk to them in high-pitched, excited voices, which causes them to see us as frantic energies, adding to the unsettled feeling they already have. In nature, the mother dog will remain relaxed to help her puppy feel calm. However, when the puppy experiences hectic or stressful energy, it may respond in an agitated or frantic way, often giving the human the impression that this is a puppy with behaviour issues. The way a human deals with certain problems is different from how an animal will deal with the same problem.

Many people forget to fulfil the dog's needs, and they end up fulfilling their own needs where the dog is concerned. The human tends to look at the dog with human eyes. It is important not to reflect our needs onto our animal friends. It is vital to understand how a dog's mind works and how they see and experience the world. A good telepathic connection can help you know your dogs on a much deeper level.

Now let's take a further look at the Canine as a species.

How often has your dog come to you while you were busy and tried to demand a game? This has often happened to me and even my clients and friends. Our response is usually to say, "I am busy, I can't play now. I will come and play with you later. Be good and go away so that I can finish working."

Dogs mirror us, so if this happens, take a look at yourself. Are you working too hard? Do you need to take a break? Do you need to play? Have you become too serious? Demanding a game is a dog's way of getting us to respect ourselves, lighten up and play more or just take a rest before we have a physical breakdown from overwork. It is often not about your dog wanting to play.

Often our animal companions are more aware of the state of our health than we are ourselves. They pick up all our emotional stress and sometimes even our physical ailments. Our Canine companions spend a lot of time trying to show us how to enjoy even the smallest things in life. They are experts at teaching us about unconditional love and acceptance.

Something as irritating as a dog that barks persistently is relatively easy to control. However, most of the time, a person will react the same way that the dog is reacting. They will go outside and shout for the dog to stop barking. The dog will see this action as his human barking with him. In his eyes, you are helping him to bark. There are some simple reasons why a dog would bark like this, and pure frustration resulting from total boredom is most likely. In some cases, some dogs are scared of the dark.

All animals need a certain amount of mental stimulation and physical exercise. Imagine being extremely intelligent and spending days, weeks or even months in one garden with nothing to do. As I mentioned earlier, our animals also mirror us. They reflect exactly what is going on in our own lives. So when you notice your dog looking restless, bored, or frustrated, take a look at yourself and see whether you are feeling restless, bored, and frustrated. You can then approach your dog with calmness, clearness, and consistency. No dog will accept an unstable leader, so it is up to us to be the leaders that our dogs can feel safe with and be proud of.

Live in the moment, accept everything around you, don't judge others, and love unconditionally. These are just some of the messages that I have repeatedly received from dogs and other animals around the world. However, what does all this mean?

Living in the moment is not about ignoring the future and forgetting about the past. It means that we should be a part of the world we live in, experience the smells of the tree bark and flowers around us, enjoy the feel of the wind on our skin, hear the sounds of the birds in the trees and feel the grass beneath our feet. We take most of these things for granted but living in the moment means never taking anything for granted, appreciating everything around us, and feeling the universe on an intense level. By doing this, we will start to access all our senses and become far more functional as human beings, rather than walking around with our eyes closed.

Accepting everything around us without judgement is about acknowledging everything and everyone around us. See all living things, even the plants, as unique beings with an important purpose. It's about allowing others to be who they are without judgement or criticism.

Loving unconditionally is not as easy as it sounds. To love unconditionally means to allow someone else to make their own mistakes so that they can learn from them. The same is true for our animal friends. As long as they are not endangering themselves, we need to give them the space to grow and learn and also make mistakes. It is not about protecting them by keeping them caged or indoors. Every animal needs the feeling of the grass under their paws and the sun on their backs. Unconditional love is not about smothering them to protect them but instead allowing them to walk their paths in happiness, love and safety.

This sounds easy, but it is challenging for us humans to live like this in reality. We naturally protect the ones we love. We are demanding in our wishes, and sometimes to protect those around us, we keep our eyes closed. We need to make an effort to start by opening our eyes and looking at the world around us. Let us try to experience our world the way the animals experience it. If we can learn from the animals and tap into our intuitive abilities, we will feel everything around us in a much more intense way.

Let us make a conscious decision to enjoy our lives and be happy. The animals have given us these messages for a reason. Now is the time to move forward and learn from our wise canine friends.

A VERY SPECIAL DOG

This next story is of Isis, an extraordinary dog.

I have never believed in coincidences. I always believe that things happen for a reason. Over the past few years, I have seen many litters of puppies in different places. I love to see puppies, spend time with them, and experience their joy and innocence.

Isis & Ginger

Many times people have said to me, "Wouldn't you like a puppy? I have a puppy that would fit into your family beautifully!" My answer is always the same. "No, sorry. I can't take on a puppy at this stage."

Well, things were about to change.

My beautiful dog Riff-Raff had developed a small mole on his nose. I took him to see the vet, and she suggested removing it. I made arrangements to bring him in the following week for this minor surgery. Whenever any of my animals have any form of surgery or procedure, I always stay with them until they are finished and can come home. This day was no different.

Riff-Raff had gone through for his procedure, and I was quietly sitting in the waiting room chatting to the receptionist when a lady brought in a couple of Siberian Husky/Malamute cross puppies. These little balls of fur were coming in for their vaccinations. Immediately I felt a huge pull in my heart. I asked if she had any females, and she said the one female was sitting in the car waiting to be brought in. I went straight out to the car to take a look.

One look at this little puppy, and it was love at first sight for both of us. We just knew each other. As there was such a deep bond between us, I decided right there and then that this puppy had to join my family. I made arrangements to pick her up the following week as she just needed to be a little older, and I also had to tell my other dogs about her before she arrived. My dogs needed to feel as if they had decided to bring her home to help divert any potential jealousy issues. It was one of the most challenging weeks of my life waiting to bring Isis home.

The day finally arrived, and off we went to pick up the special little lady. My mother came with me to drive while I held Isis on my lap. When we arrived at the house where the puppies were being kept, the littermates played and ran around quite confidently. Isis was timid and hiding, but the instant I had her on my lap in the car and we were on our way home, she sat up and became very perky and happy. She just knew that she was going home.

I let her out of the car when we got back but not through the gate. I wanted to give her some time to settle before introducing her to the other dogs. In a matter of minutes, Riff-Raff and Stacy created so much fuss on the other side of the fence that I let them through. Stacy, being the Alpha, went straight up to Isis and sniffed her, welcoming her to the family. When Riff-Raff came up for his introduction, Isis took one look at him and fell head over heels in love. From that moment on, she followed him everywhere and even slept with him at night. Stacy loved her but sometimes found her a little sharp, so she kept away. Riff-Raff took her under his large paw and looked after her as if she was his very own puppy.

Isis had only been with me for three weeks when I had to go to Cape

Town to run a workshop. My mother and sister stayed in my house to keep an eye on the animals, especially the new puppy. Before I left, I asked Riff-Raff to look after Isis while I was away and take her out at night to do her business in the garden so that she would not do it in the house.

When I got home after my trip, my mother told me that two or three times during the night, Riff-Raff would go upstairs, fetch Isis and take her outside to do her business. I was so proud of Riff-Raff for doing his job so well. This just proves the importance and value of telepathically communicating with your animals.

Bringing Isis into our family has reconfirmed to me how animals take on the behaviour of whoever they are the closest to. Isis has taken on literally every movement and characteristic that Riff-Raff has. She eats like him, plays like him, walks like him, and even sings like he does.

The first day I had her, I realised something different and quite special about Isis. It was brought to my attention when Rikki, the stray cat, walked in and joined our family. He was quite a dominant cat, and Ginger, the other cat, was nervous of him and tried to keep out of his way. Rikki had grown up on the street and was a tough little guy.

One day I was sitting watching television when I heard a noise. It was the two cats hissing at each other. Isis jumped up and ran to the kitchen where the cats were. I followed on her heels as I didn't want Isis to injure any of the cats. However, I needn't have worried. When I got to the kitchen, I found Isis standing between Ginger and Rikki. Each time Rikki tried to approach Ginger, Isis nudged him with her nose to keep him away. She was protecting her other best friend, Ginger. Isis stood between the cats for as long as it took them to settle down and move away from each other.

Over the next few years, I observed Isis's behaviour around the cats. Every time they looked at each other, she would slowly get up from wherever she was and stand between them until they relaxed. Eventually, whenever I was busy, and there was a little tension between the cats, all I would do was whisper, "Isis," and she would go and calm them down.

WINNIE'S MESSAGE

Every animal I connect with has a great deal to teach me. Winston crossed over some years ago, but his human family, who miss him dearly, ask me to connect and get messages from him from time to time. His family would provide me with questions I should ask him during our session.

Portrait of Winnie

It was one of those occasions, and I was preparing to connect with this wise little dog, fondly known as Winnie.

As I connected with him, the first thing he showed me was a huge pink cloud. He then showed me a mirror hanging on a wall. The picture in the mirror then changed, and I found myself looking at a sparkling pond of water. I had no idea what this meant.

The message from Winnie illustrates how important the imagination is and how meditation using your imagination can lead you to some quite amazing places.

I asked Winnie about these symbols, and this is what he said:

"If you climb through the mirror, you will be in a magical place, a world where no stress exists, a world where nothing needs to make sense. Everything is accepted as it is. A world very much like the world I am living in right now. I want you to go on a magical journey to a place where there is no time and no space. A place where everything is accepted and nothing is questioned. A place of love and trust.

"Here is a way for all humans to meet up with their animal companions who have crossed over and live in this magical place.

"Picture your animal friend on a pink cloud, then travel on this cloud through a mirror. Imagine this mirror becoming shiny clear sparkling water. Allow this water to cleanse all the negativity, stress, and anger out of your system. After this cleansing, you can connect with your animal friend. Enjoy your time with them in this magical place. Ask them questions and listen for the answers. Once you are finished, thank them, and focus on your breathing until you are feeling grounded."

His human companions asked him how he was because they missed him so much; it had been many long years that they had lived without him.

Winnie said, "To you, it has been many years, but to me, it feels like only a second ago that I was with you in the physical dimension. Time doesn't exist for me anymore. I was lucky to have lived in a home with so much love, and your love still surrounds me. I only have a short time to chat because I am quite busy."

I had to ask him to slow down. He was chatting so fast, and I couldn't write down quickly enough everything he was saying.

He said, "I am healing the earth by focusing on the 'hot spots'. These are the places around the world where there is little or no respect for people and animals, where there is a lot of abuse and anger directed towards the animals and the children. My job is to dig deep into the earth and plant crystals. These crystals will lift the energy and bring about huge changes in positivity, good health, happiness and enlightenment. I am working with a team of other dogs. Yes, we are the best diggers. We have no time to waste. This year will bring about a lot of changes. You might see them as disasters, earthquakes, floods, and other huge events. This is all part of the cleansing process."

Winnie's humans asked, "Are you still with us? A few days ago, we saw butterflies. Did you send them?"

"Yes, I am with you. The butterflies are there to remind you of the gentleness and beauty all around you, the beauty within you and the beauty of your soul. There is nothing wrong with shedding a few tears; I know they are tears of love.

"I have no intention of stopping my visits to you. We are soul mates, which means I will always be with you whenever you need me. I often spend the night on your bed watching over you. I always will.

The thing about a soul is that it can be in more than one place at a time.

"Significant changes like this are always unsettling, but this will end. It is a good thing that you have moved to Cape Town. Both of you need to give yourselves time to adjust to the energies around you. Cape Town has a powerful energy. It has the cleansing energy of the Ocean and the grounding energy of the Mountain. You will do well there, but you need to focus on precisely what you want out of your life so that you can bring it to reality. Be very careful not to focus on what you don't want in life so that you don't create these things."

The family asked, "Have you seen Kimi, the neighbour's dog? She crossed over recently."

"Yes, I have seen Kimi. I didn't know it was her, though. I haven't seen much of her. I have been so busy, and she is still meeting and playing with all her old friends. She is a gentle soul, a real lady. I am looking forward to spending more time with her; she is a lot of fun to be around. When the time comes for her to work, I am sure she will be working on bringing joy and fun into the hearts of animals and people who have lost theirs. She will help them be happy again. She says that her mom shouldn't be sad as Kimi is still with her. She says loss is purely a physical emotion."

Meggie was one of the family dogs that crossed over before Winnie, and the family wanted to know, "Is Meggie with you, and how is she?"

"Yes, Meggie is still with me and is fine. She is helping me plant crystals. She loves her job and is very proud."

Winnie paused, sensing that the family needed support, and then said, "You know that if you ask, I will always be there. I always am. Remember, life is about growth. The difficulties you experience are often the most significant development opportunities. If life were still and quiet, you would be marching on the spot, and there would be little or no growth. Friction needs to be balanced by love and respect. In life, everything needs a balance. Life is about sharing and respecting everything around you, including yourself. It's all about sharing. Value yourself, and everyone around you will value you. Remember the mirror. Be true to yourself."

"What do we do about the difficulties we are facing right now?" the family asked.

"This will ease but remember what I said; difficulties will bring about huge growth. How you handle these difficulties will make the difference in how your life progresses and how you settle and start reaping the benefits and bringing stability into your world. Remember, I love you and will always be there for you."

Winnie is one amazing little dog. He summed up everything without the detail in the first few moments. The pink cloud represents love, the mirror mirrors emotions, and the pond speaks to the cleansing and moving of energy and the ups and downs of everyday lives.

THE FIGHTING DOGS

Sue contacted me after hearing an interview I had given on 702 Talk Radio.

She had eight Staffordshire Terriers who had suddenly started fighting. I went to her house the following week to engage with the dogs. We lined them up along the wall in the lounge sitting on a mattress on the floor.

The story they told me was that their human mom had recently fallen ill, and there was a lot of tension in the home as a result. Each time the dogs were together with Sue, they could feel her anxiety and would start bristling. This made Sue nervous, and she would clench her jaw and tense her fists thinking, "I hope they don't fight." Well, her feelings of fear made the dogs start fighting.

I asked all the dogs to be more sensitive to Sue and her condition and be gentle with each other. We had a long chat where I told them exactly what was expected of them and that this aggressive behaviour was not helping anyone, especially not Sue. They all needed to hear me and become more relaxed and considerate.

After listening to the dogs and explaining everything that I had felt coming from them, I told Sue how she needed to be less anxious around them. She then told me that she had recently been diagnosed with cancer and was quite stressed as a result.

Three months later I saw Sue at the shop. She informed me that her dogs had listened to everything I had said to them the day I visited. She told me that they hadn't had another fight since. I told her that no animal communicator is that good that the dogs would immediately listen to what they were told. I added that although the dogs had a better understanding of what was expected of them, the main reason they hadn't had another fight was that she had changed her perception of the dogs and was much calmer around them, and that made the biggest difference.

Remember, our power lies not so much in what we say but more in what we do.

OLIVER

Oliver was a strong and wise dog. He had been contending with serious health issues for most of his life.

This is a poem that Laverne received from Oliver while she was connecting with him for his family. I wanted to honour this amazing dog by including his poem in this book.

Oliver

"I am tired but not yet ready to cross.
I will hang in till my body won't anymore
Look to the stars – they twinkle with inspiration
Look into my eyes, and you see that twinkling too
In the fading of life, there is new birth, too
As under the dead leaves of trees, the new shoots will grow
We are ever together

And while this is my Winter, with you by my side
My Spring is around the corner – with you by my side!

Sometimes I am tired and seem to give up
But then I find strength for a little while more

My time is not yet
But soon it will be
Yet forever together
Through it all, you will see."

— CHANNELLED BY LAVERNE HYMAN

A few months after Oliver relayed this poem, he crossed over. He is now living the perfect life amongst the stars, the wind, the sun, and the moon. Oliver lives on in all our hearts. He was an exceptional dog, and now he is an exceptional soul.

BIGGLES AND FAMILY

Hayley is a friend of mine and an animal communicator and healer. I asked her if she would mind if I added this story from one of her client's animals named Biggles in my book.

Hayley's client shares this story:

We are the proud parents of five Basset Hounds and one "Pedigreed

Pavement Pooch" who is convinced she is a hound too. We raised Biggles, the Basset, from a puppy but adopted the others as rescues as they were led into our lives.

About a year before, Biggles began showing unusual symptoms that started us off on a long and frustrating path of numerous vet visits, specialists, treatments and medicines – all to no avail. Then through a long-lost childhood friend, I was told about Hayley, whom she said was a lady who speaks with animals.

With a hint of trepidation, I contacted her and asked her to chat with Biggles. Hayley agreed to connect with Biggles, and the response was terrific. Biggles poured her heart out to Hayley, explaining how she felt, why she felt that way, what she needed and what we could do to help her. We were overjoyed with this interaction and feedback and immediately saw a significant change in Biggles.

We then included Boo the Basset, her brother from the same litter, into a later chat. He also had some interesting things to tell us and became much more peaceful. I believe it was the first time in his life that he had felt understood.

Hayley and Biggles connected fairly regularly, and through this, we were kept abreast of developments in our dogs' lives. What peace of mind this gave us. We also found out that Biggles was the communicator in the pack, and it then became so much easier to help them understand. We would tell Biggles if we were going out or away, when we'd be back, or any other news, and she would relay it to the others. This brought about peace amongst the dogs, especially for those who had been abandoned before we took them in and felt anxious when we left the house.

Many months later, our little twenty-one-month-old grandson became very ill with a rare blood disorder. Once again, I contacted Hayley and asked her to link up with him. It was shocking to discover just how sick, sad, and weary this little boy was; he was close to giving up on life. With much patience and determination, he was pulled back again into the light by Hayley. My dogs were extra sensitive to our angst, and I believe they all stepped in to help carry us during this stressful time.

For my husband and me, this was a turning point in our lives. We felt distinctly that our spiritual eyes, ears, and beings were opened and felt drawn to look further into the holistic way of life. We are incorporating this newfound peace and joy with wonderful results and have begun to attend courses, healings, and meditations. It's almost a sense of coming home.

In all of this, we will never forget that it was through our dogs that this new way opened up for us. If it hadn't been for them, we would not have arrived at this amazing place. We realise, yet again, how much we have learnt from our dogs and what a huge part they play in our physical, emotional, and spiritual path.

And so, to each one of our extraordinary dogs – Thank you for all your loyalty, unfailing love, and patience while we're learning lessons from you.

RIFF-RAFF AND THE VISUALIZATION TECHNIQUE

A Siberian Husky is an unusual kind of dog. They don't always respond in the way a normal domestic dog will respond. For instance, a Husky will not always come when called; they are incredibly independent and have their own agenda, which doesn't always fit in with ours.

Riff-Raff

After years of connecting on a telepathic level with animals, I found that telepathic communication works exceptionally well with Siberian Huskies.

I am always cautious when taking Huskies out for walks because if a Husky gets off the lead, he will often run for miles. He is not running to get away from his person. He is just running to explore, which can sometimes lead to disaster.

On one particular day, some years ago, while we were at dog training and doing agility, I decided to try the visualisation technique with Riff-Raff. I was nervous about letting him off the lead until I checked all the fences to make sure he would be safe if he decided to make a run for it.

When it was his turn to do the agility course, I stood in the centre of all the obstacles and without saying a word or moving, I pictured all the obstacles in sequence then let him off his lead. With that, he started running and went over and through each obstacle as I had pictured it, without missing any along the way. As soon as he had finished, he came straight back and sat in front of me, waiting for his treat. I was amazed. However, I do understand that he responded well on this day and might not respond as well the next time.

Even though an animal responds to the pictures we send, we need to remember that, like with Riff-Raff, any animal has its own mind, so it might not always respond in the way we expect. On this particular day, he responded in the way I was hoping he would.

It is important to remember that just because we can communicate telepathically with animals, it doesn't mean that they will always listen to us, but they will always hear us.

SCEPTICAL

This next story proves how animals pick up thoughts and behaviours from their human companions.

I received a call from an anxious lady having problems with her two dogs. She lived with a Weimaraner and a Pit Bull. Over the past few months, both these dogs had become vicious and unmanageable. It had become impossible for her to control them. They had started biting the gardener and not allowing any visitors into the yard. She had taken them to numerous vets and the best animal behaviourists with no luck.

I was their last chance before she decided whether or not to have them euthanised.

To me, having two perfectly healthy dogs put down is not an option. Aggression usually stems from either fear or pain. My job is to find out from the dog where the aggression originates. Once you can determine the source, you may then start the process of aiding the dog into a more relaxed and happy state.

After chatting with the lady for a long while, we set up an appointment for me to go to her house and communicate with her two dogs. She lived about an hour and a half from me, and when I finally arrived, she expressed concern about her dogs biting me and asked me where I wanted to do the consultation. I explained that I needed to work where the dogs felt most relaxed because they wouldn't tell me anything if they weren't. We eventually decided to have our conversation in the kitchen.

There was a corner table with chairs like a breakfast nook, and I sat down there. I felt sitting down would help the dogs feel safe with me and not be threatened in any way. Both dogs were a little suspicious of me but seemed relaxed.

I started connecting with them and introduced myself, explaining why I was there. I told the dogs that I needed to find out why they were biting people and behaving aggressively. I waited for an answer, but nothing was forthcoming. I tried again, and still, they said nothing. Eventually, I told them that I was their last chance, and their lives depended on what they told me. I said that I had come very far and needed to charge their human for this consultation, and I couldn't do so if they didn't give me any information.

After the third attempt, I suddenly heard the words, "Why should we tell you? She won't listen anyway. It will just be a waste of time."

They were right. This lady was sceptical of me communicating with her dogs, which made them sceptical too. The lady always had so much to say that she had no time to listen to anyone else's point of view, including her dogs.

After chatting with the two amazing dogs, I eventually got some information from them. They told me that the lady was so busy she didn't have time for them. They also mentioned there were building works going on and strange people coming into the house at all times. Neither of the dogs liked the noise or the people. They just wanted to spend more time at their farm near the dam to go for long walks and play. City life was not for them; there was too much noise and chaos.

I thanked both dogs for giving me this information and said that I would tell the lady to make the necessary changes for them to be happier.

I told the lady that the dogs initially didn't want to talk to me because they said they didn't feel she would listen and I'd be wasting their time. She started to interrupt me, so I cut her off and said that she needed to allow me to finish. She needed to listen to the information from her dogs; otherwise, they would become even more difficult to handle as they already felt ignored and pushed aside. I told her they had spoken about spending more time on the farm near the dam; this made her listen.

She then told me about a farm they had. Every weekend she and her husband would take the dogs to the farm next to Hartbeespoort Dam. She confirmed that the dogs loved it there. She also confirmed they had recently started doing building alterations and the dogs were very irritated by the constant noise.

All I could do now was wait and see if this lady would listen and do what was necessary to curb the aggression of her two dogs.

I later found out that soon after I left that day, she started making plans to take the dogs up to the farm to live there permanently. She had a very responsible caretaker who would look after them, and she and her husband would go to see the dogs every weekend.

The two dogs are now content and relaxed. They are staying on the farm just as they had asked and living the life they wanted. I wonder, are they still sceptical, or do they now believe that there are people out there that can talk to animals and other people that will listen?

STORY FROM MICKY

Micky attended one of my Animal Telepathic Communication Courses and immediately put into practice what he had learnt, with surprising results.

Micky writes:

I had received the notes for the Animal Telepathic Communication Course from Jenny. I want to share some incidents that have happened after reading and implementing some of the techniques.

I've always felt that animals and insects understand, although they don't always listen or respond. Spiders tend to climb into the glass I hold for them so that I can take them outside. A bee once flew into a cobweb when I was trying to usher him out the window and when the spider rushed down to claim his meal, I shouted 'no', and it backed off. I got the bee out of the web, onto my hand, and he sat perfectly still while I picked the cobwebs off him with tweezers before he flew out the window. Jenny's course showed me that animals and insects definitely understand.

After attending Jenny's Animal Telepathic Communication Course, my Sunday started with a bang and a lot of barking. There was a family fun run right past my house, and the dogs were excited and super-loud. I staggered out of bed and went into the kitchen, where they sleep.

One of our dogs, Tobi, has the unfortunate habit of using the lounge as her bathroom when she's allowed into the house, so they've been relegated to the kitchen and backyard. However, I'm hoping that will change now that I am using some telepathic techniques on her.

Entering the kitchen, I was greeted by excitement. I told the girls that if they were quiet and calm, I would take them to the front yard, and we could all sit and watch the people and dogs running past. They settled down almost immediately and followed me calmly to the front door.

We spent an hour sitting together on the steps, watching everyone go past, some with dogs. Tobi got a bit aggressive because of the other dogs, and I had to reiterate a couple of times that I was the Alpha, and it was okay; they were safe. The other dogs were just walking past and posed no threat, so she had to let them go. They had their breakfast out front so that they could continue to watch the procession, and when I needed to come in, they went straight out to the back without me needing to ask.

I decided to bathe them as well, and instead of trying to jump out of the bath like they usually do, I was able to wash both of them by myself. They stood quietly, if not entirely happy, while I praised them and thanked them for being so good.

Lady especially seems to respond to telepathy instead of vocal cues. I've always struggled to get her to sit and listen to me. On the other hand, Tobi has always sat as soon as I've held up my hand. It's the complete opposite with communication. Lady responds immediately, Tobi waits for vocal or visual cues.

The cats have also started to respond to telepathic communication, and it's a lot of fun for me just to sit and chat with them. We have five cats, two of which are fosters who, after a while, became permanent members of our family.

We had been having issues with Tigger bullying Munchkin. During our chat, it transpired that while we were trying to find homes for them, and because Tigger still feels like my partner doesn't like her, she has to secure her place by making sure Munchkin doesn't intrude. I tried to let her know she was safe and loved and wanted, and so was Munchkin, but she was still unsure, so we will be working on that.

I had a chat with one of my friends who is a game ranger in Limpopo. I told her about the course and how excited I was about it. Being sceptical, she jokingly asked me how I would find leopards on game drives. I told her I would try to communicate with the leopard, but I wasn't sure it would work and couldn't promise anything. She laughed it off and left me to it.

I sat down and reached out to speak to the leopard. I told her that I understood that people are irritating, and although it was difficult to explain why they wanted to see her, the people are glad when they get a sighting. I explained that she had such a lovely, safe territory because people like to see her in her natural environment but that those who keep her safe in that territory would like to see her more often. I said that if she could do that, her habitat would be safe longer as people would know she was happy there.

I got quite dizzy as I was sending the message, and for a short time, it felt like I was on the reserve, face to face with the leopard. I felt an acknowledgement and then a dismissal, so I thanked the leopard for her time and left. I had no idea whether she would listen to my message, and I felt that it was entirely her choice.

My ego wanted her to listen to me just so that my friend could see it was real and not just imagined. More importantly, I felt that maybe if people like my friend could see that animals do listen and understand, deeper awareness and acceptance of their nature would develop and make for more content people and animals.

That night I received a message from my friend. She was 'freaked out and amazed – they had seen the leopard. She was lying in plain sight, on a rock near the road, in an area of the reserve where they had never seen leopard before. My friend posted the photo on Facebook and thanked me. I told her that it had nothing to do with me; I was just passing on the message. It was the leopard who chose to listen.

There will always be a part of me that gets a thrill from knowing that the leopard did hear me and felt that my message was sincere and was happy to respond to it. I got the impression that there will be many more leopard sightings on the reserve now, for as long as the leopard feels comfortable showing herself. Hopefully, the humans visiting her space will respect her.

CHAPTER 4
Some Equine Magic

As with all the animals one wishes to communicate with, the first thing is to understand the equine species. How they think, react to certain situations, and see us humans – and that is only the beginning. There is far more to a horse than what we see physically when they are grazing in the paddock.

These animals are highly intelligent, extremely powerful and very sensitive. Only once we have made the ultimate connection with a horse do we realise that we are not the healers and we are not the teachers. The horse is the ultimate teacher and healer. Our job is to get out of our heads, calm our minds and open our hearts so we can receive the teachings the horses are willing to share with us.

As horses are such majestic and powerful creatures, sometimes we can feel slightly intimidated or be in total awe of these magnificent animals.

When a horse looks at us, it doesn't see our faults, insecurities or fears. Yet, it does "feel" our emotions. This array of emotions can sometimes be quite confusing to the horse, as it sees directly into the deepest parts of our soul. These creatures make no judgement, and we can learn a lot from horses if we are just willing to listen with our hearts. We can start by acknowledging and seeing them as a soul entity and not purely a physical form.

Horses symbolise strength, power, freedom, movement and love. They carry deep wisdom, and there is also an element of mystery and mythical magic about them.

Horses are natural-born healers and thrive on "helping" us heal in their particular way. I use the word "helping" as opposed to "healing" because they will put us on the right path and introduce us to our soul so that we can balance our own energies. This way, our bodies can heal themselves. In other words, they get our emotions balanced so that our energy will follow, resulting in the healing we so desperately need.

Have you ever wondered why horses are so good at working with disabled children and adults? These wonderful animals can connect with disabled people of all ages and share their love, and the person receiving this affection responds in a highly positive way.

Many years ago, I helped an entire school of mentally and physically challenged children with Animal-assisted Therapy. I had five remarkable horses at the time, ranging from two miniature horses to a huge Thoroughbred gelding and even a Boerperd. A group of twenty children would come out to the stables every week to be led around on one of the fantastic horses.

We had children with various incurable problems. One little boy had so much muscle tension and calcium build-up in his system that he couldn't bend any of his joints; he really was like a stick figure. Another young girl with no muscle control was like a rag doll and needed to be held upright on the horse as she couldn't hold herself up. Then we had the blind boy who had the most amazing connection with the horses.

We always had one person to lead the horse, and two of the teachers from the school would hold the child on either side while they were riding. The biggest challenge was managing the teachers' fear of the horses.

After a year of the children visiting the horses, we held a basic gymkhana for them. All the parents and the school principal came to watch the children riding their horses. The visitors were thrilled to see the little boy with stiff joints, whom they had nicknamed Pinocchio, sitting perfectly on his pony. The girl with no muscle tone was sitting up by herself with only one person to steady her, and the blind boy was riding on his own.

The horses had worked their magic, and all the children had dramatically improved in the space of a year.

Most humans have so many blockages that have developed over the years that they cannot see the energy fields surrounding everything. However, horses can easily see the energy. They can even see the energetic vibration coming off the earth and the plants around them.

For those of you who have engaged with horses, have you ever wondered why you battle to catch a horse some days, especially when you are in a bad mood or when you are not feeling well or are suffering from stress? This is mainly because horses respond to the energy vibrations that they see radiating from you, and this can be quite unsettling to a horse.

Even though there is a mystery surrounding horses, the concept of independence is totally alien to them and sometimes fairly stressful. They function in herds where there is respect, loyalty, love and companionship. However, the domestic horse will at times compete for food or water.

In one of my early workshops, Sue had an absolute fear of the horses. I decided to put her with one of my miniature horses named Penny; she was small, gentle and un-intimidating. I stayed with them in the stable for a while, then left them on their own so that Penny could work her magic. On my return, I found Sue sitting on the stable floor, and Penny gently resting her muzzle in Sue's neck, blowing softly.

Sue was in tears and astonished by Penny. Sue had a life-changing experience, and from that day, she no longer fears horses; she only has deep respect and love for them.

Horses are incredible to interact with, and they have so many lessons to teach us. All we need to do is give them the opportunity and allow them into our hearts.

Penny teaching

SOME HORSE MESSAGES

Here are a few messages from some of my special horses. These messages are taken from my book "Paws & Listen to the Voices of the Animals".

Penny says:

> *"When you look at us, open your eyes and see us.*
> *Appreciate our finer points.*
> *We are there for you.*
> *Learn from us, and we will learn from you.*
> *We must work as a team to better the world."*

Ballinor's message:

> *"We all have something to say.*
> *You need to allow us to speak and listen to what we say.*
> *You might be pleasantly surprised."*

Tootsie's wisdom:

> *"The world is there for all of us to share.*
> *We are all a part of the same Universe, and we need to realise this.*
> *For everyone*
> *Human, Animal and Plant*
> *Is here to make a difference.*
> *We are all important in the bigger scheme of things.*
> *The sooner we realise that we are all part of the same oneness,*
> *The sooner we will feel the love of our fellow beings.*
> *And that of the Universe we live in."*

Connecting and spending time with horses is truly a magical experience. The horse is a profound and spiritual being, but first, let's try to understand them on a physical level.

Horses are herd animals, and in the wild, they live in groups of between six and twelve horses. Communication with each other is done physically using body language. This happens in subtle ways – a flick of an ear, a swish of the tail, a curve of the body. If the other horse ignores or defies these signals, they resort to contact such as nudging or nipping and even biting and kicking at times. The communication stops immediately when one horse gets the desired response from the other.

With this in mind, when you train a horse, they learn that they have achieved what you want when you stop asking. The horse does not know right from wrong, nor does it make moral judgements. If the horse responds and you stop asking, the horse considers their response to be the right one. Similarly, if the horse does not respond, and you stop asking, the horse sees not responding as the correct behaviour. Horses learn by repetition, so be sure which behaviour you want to reinforce.

On a spiritual level, horses have an extremely close telepathic bond with each other and a deep connection with their souls. One of the main lessons they teach us is to connect with our own souls and discover our true purpose in life. They are particularly sensitive, and this makes horses good at relating to mentally or physically handicapped children or adults and helping them find confidence and happiness within themselves.

Humans often require animals to do things that are entirely alien to them. Some man-made machines are foreign and frightening to the horse's heightened awareness and sensitivity.

Some people don't understand why horses shy away from moving objects, like flapping cloths, or are fearful when they hear rain falling on their stable tin roof. They cannot grasp why a horse gets jumpy when they see an unfamiliar object.

The answer is simple; horses were not designed for confinement to man-made structures. Their sight enables them to see things differently from their human companions. Nature has adapted them to open spaces, and the perception of a strange movement is a signal to flee. If we see things from their viewpoint, their behaviour seems logical. It is a credit to their willingness to adapt to the many alien situations we put them in.

Observe the horse for a moment.

Horses have a powerful sense of fight-or-flight. One moment they can be grazing happily in a large paddock, and to the human eye, they look totally relaxed. Then suddenly, they will hear a noise and, without warning, be off at full gallop across the field.

Horses have incredibly strong senses and are tuned in to everything going on around them at all times, so they have the most profound ability to help people with emotional issues. As mentioned above, one of their purposes is to help us discover our own soul on a significant level and discover our true purpose in life.

Part of reaching your optimal level of communication with animals is to understand all living beings extensively.

When working telepathically with any animal species, including the horse, you don't need to be with them physically when making the connection. No distance is too great for a telepathic link. Think of it as making a long-distance phone call. You first dial the number; this is where you connect with the energy of the animal you wish to talk with. Once you have established a connection, the line is open, and your conversation can begin.

Remember that when your conversation ends, don't slam the phone down on the animal. Thank them and let them know that you are now finished but will connect with them again later if the need arises. If you are talking to a horse or any animal, and they have more to say, they will initiate another conversation by popping into your mind when you least expect it. That will be the sign to reconnect with the animal.

A HORSE AND HIS LADY

It took Amanda two years to build up the courage to ask me to come and help her with the problems she was having with her horse. I didn't quite know why it had taken her so long to ask for my help, but I was soon to find out.

My friend told me about Amanda and that she had imported a Dutch Warmblood horse into South Africa but that he was not doing well in quarantine. Amanda didn't make contact, though, so I gathered that all was well with the animal.

I was astonished when I received a call from Amanda two years later. She said her horse had become scary, and she didn't know what to do. I didn't know what "scary" meant, but when she came out to meet me at the car, her first words were, "Oh, I am so terrified."

I asked her what she was terrified of? Was she scared that her horse would tell me something she didn't like? She answered, "No, I am worried that you will bring something dark into the home."

Perhaps she thought I would harm the horse, but after I told her exactly how animal communication works and what I would be doing, she was happy for me to go ahead and talk to her horse. She immediately told me about her horse, but I stopped her abruptly and told her I wanted to find out everything from her horse first. I didn't want to give her any reason to doubt what I was picking up from her horse; every bit of information had to come directly from him.

We went down to the paddock, and I pulled up a bucket to sit on, with Amanda next to me. It took only a few seconds, and Shadrock, the horse, walked straight up to us. Amanda was astonished and said this was the first time Shadrock had freely approached anyone in the paddock; he usually kept well away from everyone. This remarkable horse stood for forty-five minutes with his muzzle on my shoulder, and he didn't leave my side for even a second while I was talking with him.

Shadrock gave me a clear picture of himself being ridden in a huge arena, with his rider wearing a bowler hat and tailcoat. It became evident that he had been a high school dressage horse. He gave me an image of his rider, a young lady, probably in her early twenties, with long blonde hair that she often wore in two plaits. He then gave me a description of a large rough-looking man. The man was harsh and insisted on Amanda riding with spurs and a whip, and being ridden like this made Shadrock fearful. He was a huge horse and, because of his fear, had started walking backwards, refusing to enter the dressage arena. He had even started to rear up on occasion, making Amanda terribly scared of him, and in turn, her fear had rubbed off on Shadrock, who had become even more afraid.

The large man that Shadrock described was Amanda's new riding instructor, and she confirmed all the information I had received from

Shadrock as accurate. While she was apprehensive at first, Amanda now understood what her horse was having trouble with and how to solve the issues.

My suggestion to Amanda was to get a new instructor, which she did, and a week later, I visited her to find that she and Shadrock had gone on a long out-ride together for the first time in two years. The communication had helped her understand Shadrock better, and it had helped Shadrock understand Amanda. Now that they understood each other and had developed a mutual respect, they could finally develop a close bond and start working well together.

It is always rewarding to see how a communication session with an animal can help both animals and humans heal and become much happier together.

A DISSATISFIED PONY

One afternoon in October 2019, I received a call from a panicked lady. Her daughter's pony suddenly became lame whenever he was saddled up for the young girl to have her riding lesson.

Judy and her daughter Emily had become quite desperate to find out why this was happening. Emily loved her pony, Flash, and was terribly worried about him. We organised a time for me to go to the stable yard to speak to Flash.

I saw a pale grey pony standing in a paddock off to one side when I arrived. Nobody else was around except Judy and Emily, which suited me because I need total silence to make a meaningful connection.

Flash was a little tense and quite suspicious of me and wanted to know who I was and what I was doing there. I introduced myself and told him that I was there to find out if there was anything wrong with him or see if there was anything specific that he needed.

Well, the flood gates opened and out poured a story from Flash. He told me he was bored and wanted more fun things to do. He didn't like his current routine because he couldn't show off his skills and was only allowed to walk and trot in circles. He described a man who came twice a week, stood in the middle of the arena shouting and making him do boring things. The only way he could get out of working was to become lame. Clever Flash learnt that the moment he became lame, they would stop his work and allow him to rest. Flash then told me that he loved Emily and had a close bond with her. He wanted to make her proud, but he wanted to do more.

I explained all this to Judy and Emily and asked them to add more fun things to Flash's routine. I also asked if they could find a different teacher that Flash would like more, which they did, and a month later, I heard from Judy that Flash was a different horse. He was enjoying his new routine, and he hadn't had any more episodes of becoming lame since the day we spoke.

This story proves that horses, and indeed other animals, can think for themselves and be quite resourceful in getting their messages across.

HEALED BY A DONKEY

I adopted Lucy from the Highveld Horse Care Unit in Meyerton. She was a youngster at the time and totally adorable. Lucy had been pulling a cart with her mother in a township when a car came speeding around the corner and crashed into them. The horse care unit was called out, and on arrival, they discovered that Lucy's mother had broken her back and had to be euthanised on the spot. This was traumatic for little Lucy, and the horse care unit took her into their care.

Lucy

On one of my visits to the unit, I found Lucy. She had been there for five days when I adopted her and finally brought her home to live with my family of horses and me. A few years after adopting her, I started getting her to help in my workshops, where we learn about telepathically communicating with animals and focussing on different healing techniques. By having her in the workshops, I discovered she was an accomplished healer.

One day, after spending some time with Lucy in her stable, I turned to leave and felt a gentle breath on my neck. Lucy had reared up behind me, and I felt her muzzle gently resting on the base of my neck. I could feel a warm energy moving down my spine. She and I stood like this for about ten minutes before she finally stood back down and put all four hooves back on the ground.

I had broken my neck many years before in a serious horse-riding accident. Lucy had sensed my pain and given me some healing to relax my stiff neck.

I am always astounded by the perceptions and abilities of the animals around us. We just need to be more aware of what the animals are trying to tell us. We could learn a lot from them.

REBECCA'S LESSON

When people join a workshop to learn how to talk telepathically with animals, they often feel that they need to "hear" the animal's voice. They also think that if the animal is not standing right in front of them and making eye contact, it is not talking or listening.

On one of the workshops, I placed Sarah with Rebecca, my beautiful Thoroughbred mare. As we all started connecting with the animals we were paired with, Rebecca strolled up to Sarah, and before she could finish connecting, Rebecca walked off.

Sarah was devastated and told me

Rebecca & Jenny

that Rebecca didn't want to talk to her; what must she do? I asked Sarah to finish connecting with Rebecca and to ask her why she had just walked off. She must have a reason.

What Sarah got from Rebecca was, "Just because I walked off doesn't mean that you can't talk to me. I can hear you and can talk even while grazing in the paddock. I don't need to be right there with you to talk."

Rebecca gave Sarah many lessons that day and, in doing so, had helped Sarah develop her telepathic skills.

CHAPTER 5
Animals and Crime

Crime affects our animals in much the same way as it affects us. The animals living in our homes are sensitive and feel like an integral part of the family.

We don't recognise the effects crime has on them. Our dogs, and sometimes cats, often feel responsible for allowing the perpetrators into our homes and not being able to prevent crime from happening. Animals may even put themselves at risk to try to protect us. Many people try to stop their dogs from barking, only to discover that they are warning them of a stranger lurking on their property.

It is vital that we are sensitive and take note of the animals around us. If they seem to be unsettled or stressed in any way, it is up to us to find out why, as there is always a reason. Just as a child could need counselling after experiencing crime, an animal will often need counselling and care following an incident.

The following stories will illustrate some of the ways that crime has impacted some animals and how they were helped through this traumatic event.

STOLEN

The names of all the people and animals in the following story have been changed to protect those involved.

Early one evening in May 2009, I received a phone call. A lady, we will call her Sarah, was distraught and explained that her dog Layla had been stolen. Sarah was extremely worried that Layla might have been taken out of the country. I asked her to send me a recent photograph, and within minutes it arrived. I immediately went to work and connected with Layla.

On connecting with her, I could feel that she was unsettled, anxious and confused. I needed to know whether she was still in the country, so my first questions were, "Layla, how are you? Are you injured? Did you go on a long journey in a big metal bird in the sky?"

Most animals don't have a concept of an aeroplane and need a description to be able to understand what the term "flight on an aeroplane" actually means.

Layla told me that she had gone on a trip in a car. She had been in one before when she was taken to the vet, so this she understood. I asked her to give me as many details as she could remember.

She told me that she lived in a home with two ladies. It was a happy home where she felt safe. Her special friend Maggie was a smaller dog, and Layla felt it was her duty to protect and look after Maggie.

Layla and Maggie had felt a little tension in the home over the past few weeks, and one morning there had been a massive row between Sarah and her friend Helen. During the course of this argument, Helen had packed a suitcase, taken Layla, and left in the car. Layla was very close to Sarah, and Helen had taken Layla away out of spite.

Just like in a custody battle, the ones that generally suffer the most are the children, but in this case, it was Layla and Maggie. They were confused and scared. Why had they been separated? What had they done? I explained to Layla that she and Maggie had nothing to do with the argument; it was a human thing and certainly not their fault.

I asked her to give me a description of the house where she was now. The picture she showed me was of an old house with a small garden. There were roses along the fence and a small white wooden gate. The road just outside the white fence was busy and quite noisy. It was a Cape Dutch style house with slightly shabby walls that needed some paint. There was a large wooden door.

I asked Layla to open the door and take me inside. Layla proceeded to give me a vivid description of the house, from the wooden floorboards and tall standing lamp with a maroon lampshade to the small floral pattern on the upholstery. There were two dog baskets in one corner of the white-tiled kitchen with grey blankets in them. One had a red plastic ball in the middle.

I asked her if she knew where this house was? She said the name of the area sounded like "Orange", but she wasn't sure. She missed her friend and wanted to go home.

On contacting Sarah to tell her about what Layla had described, she told me that the depiction was of her house, the place from which Layla had been taken. It was in Orange Grove, and the description was totally accurate. She also confirmed the argument between her and Helen and that Helen had taken Layla and driven off. Helen had wanted retribution, and the best way she could think of doing this was to take away the one thing that Sarah loved the most. She didn't for one minute stop to think of the damage that she was doing to both Layla and Maggie, the innocent dogs in the middle of this battle.

The police were called in to investigate. It took four long months, but with Layla's detailed descriptions, she was finally found by the police at Helen's new house in Cresta and safely returned to her home and her friend Maggie.

It is so important for everyone to understand that the animals we share our homes with are far more than just animals; they are living breathing souls with feelings, thoughts, likes, dislikes, and friends of their own. They develop extremely close relationships with their humans and other animal companions. For their humans to go through a divorce and separate them is exactly the same as separating children – it can leave lasting effects and be extremely damaging and stressful. They should never be used as pawns in a human argument.

ARMED ROBBERY

In June 2018, I was invited to the home of a young family who had just experienced an armed robbery and were worried about their dog, Patches.

Patches, a stunning black & white bull terrier, was a much-loved family pet. The family grew concerned about Patches following a break-in at their home in which guns were involved. I was asked to chat with Patches and let him know that it was not his fault and that he could relax as he and the rest of the family were now safe.

On arriving at their home, I saw that Patches was sleeping under a table where the family was gathered in the corner of a large kitchen. I asked them not to give me any information about what had happened as I wanted to get the information directly from Patches.

Little did this family know about the information that their amazing dog was about to give me.

The minute I connected with him, a movie started playing in my head. I saw the picture of two men pushing into this kitchen, both of them armed. One was a tall thin man with very bad skin wearing a light green cap with dirty blue overalls and shabby takkies. The man had threatened to shoot the dogs, but the husband had pleaded with him and put the Patches outside on the patio while the little Maltese had run away and hidden under one of the beds.

After Patches had been safely locked outside, the suspects proceeded to tie up the family with belts they had found in a cupboard. The robbers then ransacked the house and left out the back door. Patches told me that there were actually three men involved. One had stayed outside next to a scruffy looking bakkie with a broken headlight and mud all over it. He gave me a clear description of this suspect, who he said was older than the other two and was quite short and stocky with a thick mop of black hair and an unshaven face.

Patches told me that even though he could see everything going on inside, he had felt powerless, wanting to help but locked outside. I explained to him that his family was relieved that he was unhurt and that if he had stayed and tried to help, he would undoubtedly have been shot, which would have been disastrous for everyone. I said that all that mattered was that he was safe.

When I had finished chatting with Patches, I went through the notes I had written and recounted to the family what Patches had seen and told me. They confirmed everything but didn't know that there was another suspect waiting outside. I asked them to take the description of the third man to the police after I left. Three days later, the suspects matching Patches' description were apprehended, and some of the stolen goods were found in their possession.

In a case like this, you can never tell the police that these descriptions were given to you by the family dog. All you can do is tell them that this is a description of the suspects and this is the vehicle they were driving. The police need to follow up on all the leads and make the necessary arrests.

Later I was told that Patches was back to normal. Speaking with him again, Patches said he was happy that he had made a difference by giving the information that ultimately helped in the apprehension of these suspects. The little Maltese was also reported to be safe and sound.

HOUSEBREAKING

On another occasion, I was asked to visit the home of an elderly couple who had been broken into one night while they were out visiting friends. The couple had two elderly Great Dane dogs. Once again, I asked not to be given any information as this can hamper the telepathic information I receive from the dogs. Sometimes having knowledge of the situation can be misleading, and you never really know if the information you are getting is accurate, or rather something that you might be thinking about while connecting with the dogs.

Neither of these two dogs was stressed, and both were at ease when I arrived. However, as I sat down with them, the older of the two, Bella, immediately started telling me about the "incident". She thought of it as an incident rather than a housebreaking, and I asked her why, as someone had clearly stolen items from their house. Bella told me she knew who it was and proceeded to show me a picture of a domestic worker who was middle-aged and a little plump.

After she had given me a full description, she told me that this lady had taken them and shut them in the bathroom. She then showed me a picture of a short young man with lighter black skin and a high-pitched voice. This young man had been the one who had stolen the radio and television, as well as some clothes from the bedroom.

Bella said that she hadn't seen this lady for a few weeks, so she didn't know why they had come in now. It turned out that the lady was the domestic worker who had been released of her duties a few weeks earlier, and the young man was her son. The dogs knew her, and so they went quite willingly to the bathroom with her while her son came in and stole the goods from this elderly couple.

MURDER

One cold winter morning, the phone rang. Answering it, I heard the panic-stricken voice of Hilda, a long-time client of mine. I asked her what was wrong, and she responded, "They have just found the body of a very good friend of mine, and it looks like she has been murdered."

She went on to tell me that her friend Sandra had been found strangled in her bedroom, and her little Jack Russell dog was found hiding under the bed. I told Hilda that I would go straight over to the house.

When I arrived a short while later, Hilda was sitting in her lounge with the little Jack Russel terrier named Wendy. She said she was going to be looking after the little dog until they could decide what else to do.

I asked Hilda if I could chat with Wendy to see if I could get any information to shed some light on what had happened.

When dealing with a traumatic event, I never ask the animals to give me specific information. For them to give me accurate information, they need to re-live the incident, and this can be extremely stressful for them. I usually just give them some space and see what they are happy to share on their own. Usually, this method will give me the same information but will cause less stress for the animal concerned. Also, when working with animals and crime, you must remember that nothing you get from an animal will ever hold up in court. An animal might give you certain pictures, feelings or thoughts, but this does not mean that the information they are showing you is linked to the suspects. Such information could be influenced by what the animal is thinking about at the time, and it might not have anything to do with the crime. All I can do as an outsider is pass this information on to the family, who can then get the police to investigate.

The information I got from Wendy was that there had been a huge fight with lots of shouting and pushing around. Terrified, she had hidden under the bed. I asked her if she knew who had been fighting with Sandra. She told me it was Sandra's boyfriend and proceeded to give me a strong smell of aftershave. It smelt like Old Spice. Together with this, I sensed the smell of spicy food and motor oil. I asked Wendy what else she noticed, and she said there was a car parked outside. She showed me a picture of a light blue Golf with the beginning of the registration number TWX.

Interpreting this information, it seemed as if the boyfriend had committed this murder. He and Sandra had a fight, and he had grabbed the cord of her dressing gown, strangling her while Wendy hid under the bed. However, on later investigation, it was found that her boyfriend had visited her, they did have a nasty fight, but he had left in a temper. A short while after he had left, an intruder had broken in and committed the murder. In the dog's mind, however, all the details ran into each other, and because she was so afraid, she had misinterpreted what had actually happened.

When working with animals, information can be confusing, mainly because they don't see things the way we see them. When animals relay certain information, it is often up to the person communicating with them to connect the dots and make sense of it before passing it on to anyone else.

THE GUARD FISH

This next short story shows us exactly how observant and aware all animals are.

Late one night, Michael was woken up by the sound of his fish banging against the tank and plopping up in the water, making a great noise. He lay in bed a short while, wondering what was disturbing them.

Eventually, Michael got up and went through into the lounge, where he was confronted by an intruder in his house. Luckily for Michael, the intruder ran off the minute he walked in. The fish knew it was the middle of the night when no one was supposed to be around. Not recognising the intruder, the fish had alerted Michael the only way they knew how, working together to make as much noise as they could. The dog was fast asleep on Michael's bed, so the fish were on guard duty that night.

This just shows you exactly how aware the animals around us really are. Who would ever have thought that a fish could be aware of a stranger in their home?

CHAPTER 6
Some Other Animals Speak Out

THE PARROT

On a visit to Frank, my IT technician, I was sitting in the dining room while he worked on my computer in his study. In the corner of the dining room was a cage with a beautiful African Grey Parrot. I decided to talk with this bird while waiting for Frank to fix my computer. Connecting telepathically, Frank couldn't hear what I said to the parrot or what it said to me.

When Frank returned about an hour later, I asked him who the tall, slim woman with shoulder-length light brown hair was. Her name was Susan, and I told him that she used to come here often but hadn't been here in a while.

Explaining that Susan was his ex-wife, Frank wanted to know how I knew about her. I told him that his parrot had told me. Frank was intrigued and wanted to know what else it had told me. I just laughed and said to him that his parrot was missing her visits.

This remarkable African Grey had given me a complete description of Susan, even what she was wearing on her last visit. I am constantly astounded at the details that some animals give me about their humans or other people in their lives.

A VERY CLEVER AND BOSSY PIG

Pigs are such intelligent, sensitive, and playful animals with a good sense of humour. They can also be highly spiritual beings with deep wisdom, and it is always rewarding to connect with a pig on a telepathic level.

Jafta & Merri

Pigs are clean, hygienic animals. The only reason we see them as dirty is that they love to roll in the mud and are often kept in a pigpen, so they have nowhere else to "do their business".

My two little potbellied pigs, Merri and Pippin, had the run of the property; however, every time they needed to do their business, they would do it in the same spot on the property every time. It made it easy for us to clean up and keep the stable and their roaming and playing areas clean.

For many years, Jafta, one of my gardeners and stable hands, had been helping me look after Merri and Pippin, and he became very close to them.

Jafta tells this story:

Early one morning, I woke to hear someone knocking on my door. Opening it, I found Merri, one of the potbellied pigs, waiting for me. Merri was wearing her hungry face, and she said, "Hey Jafta, don't you see that I am hungry and angry."

I didn't answer but got up and went to fetch her food. She said, "Jafta, I'm warning you, if you are going to feed me late, I will tell on you."

After eating her food, she said, "Please, Jafta, I don't like to eat my breakfast late."

I said, "I'm sorry, Merri, it won't happen again."

We carried on communicating, so I asked Merri a few questions.

"Merri, how do you feel living here with the dogs, horses, cats, rats and donkeys?"

"I feel good because we don't eat the same food, and we don't share rooms."

"Merri, how is your health?"

"I'm feeling good, as you can see."

"Do we treat you well?"

"Very well, Jafta, you are always here for us."

After talking to Merri, she gave me some advice about my life.

"Jafta, trust yourself by trusting others; there is something greater than yourself."

I have learnt perseverance and how to communicate with other animals from Merri. All thanks to her for working with me.

THE SHEEP JOINED THE CONVERSATION

A friend of mine was worried about his horse. She had become skinny and weak, and he didn't know if the time was getting close for her to cross over. I agreed to go to Walkerville and chat with her to find out how she was feeling and if there was anything she needed.

When I got there, I found her and my friend waiting in a large paddock, seemingly by themselves. The other horses had moved to the other end of the paddock, so we had quiet with no disturbances during our conversation.

As I stood waiting for the horse to start the conversation, I heard many questions, "Who is she? What does she want? Why is she here? Is she also going to chat with us? What will we say?"

I knew that these questions were not coming from the horse. I turned around to see who was standing behind me, only to find a flock of inquisitive sheep in the next-door paddock. I explained that I was there to talk with the horse as she was not well, and they must please not interrupt us. The sheep immediately stopped talking so we could go ahead with our conversation.

The horse then explained that she needed a change in diet, as well as a warmer blanket. She had recently lost weight and was feeling the cold. She also needed more bedding in her stable. I relayed her message to her human, and the changes were made straight away. Once these changes had been made, she started to slowly recover and, after a few months, became a happy and healthy horse again. On the other hand, the sheep were amazed that I would come over and talk to the horse and not pay much attention to them.

This story reminds me that we should never underestimate the animals around us. They all have something to say and should always be given the opportunity to speak and be heard, just as the sheep had told me. Sheep have so much intelligence and wisdom, and you will never know just how special they can be unless you connect with them on a deep level. Although sheep follow each other and stay in flocks and each has its own personality and can be quite entertaining. They teach us the importance of group energy, working together, and not always trying to do everything independently. They show us how to value those around us.

BIRD OF PREY

When connecting telepathically with any animal, it is not necessarily about getting heaps of information. Sometimes all it takes is one sentence that makes the difference, leading to a well-balanced and happy animal.

During a visit to a bird of prey centre, I was asked to find out why one of the eagles had suddenly become aggressive. After getting on with and loving his handler, the eagle suddenly attacked and wounded her during a demonstration of his flying and returning to handler skills.

When I asked him about this, he mentioned that there had been a change that disturbed him. His handler suddenly looked different, and the eagle didn't like it. When I asked her, she told me that she had recently cut her hair and changed her look.

I explained to the eagle that if he had his feathers cut, he would still be the same bird, and so because his handler had cut her hair and changed her look, this didn't mean that she was any different to what he was used to. She was still the same person. It took him a day or two to accept the change before once again becoming happy and settled.

This just shows that even the smallest thing can make the difference between being calm or becoming aggressive. It is all in the understanding of how animals relate to different stimuli.

It is essential to be totally open and sensitive to the information coming through when working with animals. Telepathically connecting with animals is one long lesson, and we need to be prepared to receive the information that comes through during our conversations without any judgement. The secret here is to open our hearts and get out of our heads. If we can do this, then we will make significant progress.

GECKO

A few years ago, I ran an animal telepathic communication workshop at a wildlife resort. It was a tranquil Sunday, and Sandy, my business partner at the time, and I were staying in a cabin right on the banks of a beautiful dam with a resident hippo. There were elephants, zebra and even a few wildebeest that strolled past while we were working.

On this particular morning, Sandy had gone up to the lodge to fetch the workshop participants while I stayed on our veranda to meditate and prepare the animals for this workshop and let them know exactly what we would be doing during the day. While I was deep in meditation, I heard a little gruff voice. I couldn't hear the actual words and thought it must be some builders returning as they had been building some chalets on the other side of the dam. Eventually, I made out the words, "You never remember the little ones."

When I opened my eyes, there was a little gecko in front of me on the rail of the veranda. He went on to tell me that whenever we did workshops like this, we always focused on the bigger animals and forgot that the smaller ones had something profound to say and also had lessons to teach us. I promised him that we would do a special gecko meditation just for him during the workshop. We did this acknowledging him and all the other little animals and insects. One of the messages that came from the gecko was that even though she is small, she still carries a huge amount of wisdom and power in her soul.

After that day, I never forgot to connect with and get messages from the little animals and insects who also have so much to teach us.

Thank you, Gecko, for the reminder.

THE SWALLOW

One early morning while sitting outside, I noticed a whole flock of swallows sitting on the phone lines at the bottom of my garden. They were getting ready to migrate, so I decided to get a message from them before they flew away for the winter.

They gave me this wonderful poem.

JENNY SHONE

"With one body in unison, we go
Soaring high in the sky, we fly
Looking down on the earth below
Not a worry, not a care we feel
We have purpose and a mission to achieve

Without judgement or pain or fear
We are ready to depart this year
It is hard, it is tough, we know
But we have a mission and a place to go

Soaring high in the sky, we fly
Looking down on the earth below
The breath of the wind helps us go
We are safe, we are happy, you should know

The trees whistle, flowers sing, mountains wave us on
*   our way*
The animals look like shining crystals down below
What a feeling, what an honour to be high up in the sky
Looking down on the earth as we fly

One body in unison, we go
Don't worry; we'll be back, you know
We are happy to be flying in the sky
Looking down on the earth below
We have a mission and a place to go
We are known as
The Swallow"

— CHANNELLED BY JENNY SHONE

A SWALLOW BLESSING

Over the past few years, the swallows have lived in three swallow nests right above my front door. One day, I noticed that a pair was scouting for a spot to build another nest. They decided on building it just above the glass sliding doors on my patio.

It surprised me that they decided to build there as it was noisy, with people coming and going and dogs and cats also living there. However, this is where they decided to build.

I welcomed them and told them to please be careful of the cats and the dogs. I told them how happy I was to have them where I could watch them come and go as they built their nest. I could also watch them hunt every time the mosquitos and insects flew around.

After three attempts to build the perfect nest, they finally settled. A few weeks went by, and then I noticed there were four swallows. They had babies who were now flying around with their parents. What joy this gave me.

I was having my cup of tea early one morning; the sliding doors were open, cats were asleep, and the dogs were lying outside on the cool grass. I greeted the swallows as usual when suddenly, they flew into my lounge through the open door. They flew about a metre above my head around in circles for five minutes, then ducked and flew back out.

They repeated this exercise and circled above me for the second time, looking deep into my eyes as they did this. The little birds were thanking me for welcoming them into my home and respecting them the way I did. They thanked me for my energy and for sharing it with them. The swallows told me they would soon be leaving for warmer territories, but they would be back.

They showered me with silver-blue crystal-like sparkles as they left. I could genuinely say that my swallow family had blessed me in the best possible way.

A MESSAGE FROM THE OCEAN

On a visit to a friend in Plettenburg Bay, I decided to see if I could get a message from the whales that I had seen a few days earlier. I got up early one morning and went to the beach, hoping to connect with them. I was on my own when I went to a lookout spot with a perfect view. It was drizzling as I sat down on a bench, and then I saw them in the bay.

As I connected with these beautiful creatures, they disappeared back under the water. I lost sight of them, but I could feel their powerful energy. I asked the whales why every time I wanted to connect and get a message from them, they would go under the water where I couldn't see them. They told me that being able to see them would be too distracting for me. While they were connecting with me, they would be out of sight, enabling me to focus on what they were saying rather than what they were doing – a valuable lesson for us from a wise animal.

Sitting on the bench with the whales close by beneath the water, I captured the message from these great and elegant souls.

This is what they said:

"Sound carries vibration, and vibration carries energy. Our form of communication carries an energy of the highest form of enlightenment and healing on the deepest level. It forms a net around us, which expands and spreads through the entire ocean as we move. This vibration is not restricted to the sea. It travels deep under the earth across the entire planet.

We connect with all life forms through our ultrasonic sound vibration.

The bottom of the ocean is alive with energy and life, unlike the world outside. Your world. In your world, there are still many people walking around dead, simply existing from one experience to the next, never knowing what it feels like to live with all their senses awakened. Never knowing what it feels like to really be alive.

With the progress of the awakening of the energy, which is

happening right now, many souls will depart from this planet, and some will experience the light and grow in it.

We, the whales, are coming to the end of our existence on this planet; we are preparing to leave. When we go, we will not be coming back. In what you perceive as time, it will still be a long while, but it will be soon in our understanding of time.

The main lesson for anyone to learn is the lesson of respect. With respect comes acceptance; with acceptance comes love. Respect yourself, accept yourself, love yourself. Once you have accomplished this, you will respect, accept, and love all living beings around you. Only then will you awaken and start to function in the light and be alive for the very first time in your life.

This message must be shared with others. The way our ultrasonic vibration spreads through the universe is all about feeling the connection to all life forms. It is not about secrets and keeping the knowledge inside you; it is about letting it out and sharing it with others. As with energy, the knowledge will spread.

As we hold and share the energy of the sea, the elephants hold and share the energy on land. These animals carry a powerful vibration of energy and ultrasonic sound. Their energy is different from ours but just as powerful. While we share the energy of healing and enlightenment, they share the energy of wisdom and growth. With healing and enlightenment comes wisdom and growth. You can see how important it is for us to work together. We all need to work for the betterment of our planet and all who inhabit it.

Everything in life is connected. When one is hurt or suffers, it affects us all. How often have you felt depressed, ill or unhappy for no reason? Well, maybe it is not for no reason; perhaps someone on the other side of the world is hurting, and you are picking up the energy of their vibration and being affected by it.

Move forward in confidence and truth.

Go out and enjoy life.

Experience life.

Live life with your eyes wide open."

I am always so grateful to be a part of a world where animals talk and people can hear them. It's a magical world that everyone can experience. All it takes is an open heart and a clear mind to feel and hear the wise messages coming from the animals around us. Even our domestic animals have many lessons to share; we just have to be open enough to listen.

A CONFERENCE WITH A MISCHIEF OF RATS

I had become extremely frustrated with the rats living in my stables and helping themselves to the horse feed. They were taking over my large feed room, opening bags and climbing into the drums to steal and soil the horse and donkey feed.

I had tried for months to get them to move out and leave the feed alone. I even reached a compromise and told them that I would provide for them each day if they left the horse feed alone. They just did not listen. I was reaching the end of my tether. People said I should poison them, which I refused to do as rats are intelligent and sensitive animals. I don't believe in or agree with killing any animal.

One day, I decided on a plan. I placed a rubber dustbin in the feed room. The dustbin was quite deep, and it had a lid. I put some horse feed in the bottom of the bin and laid a small water container so the rats wouldn't die of hunger or thirst in the hot feed room. I then cut a small hole in the lid for easy access.

Two days later, while I was in the feed room, I heard a scratching noise coming from the dustbin. I slowly opened the lid and saw at least twenty rats sitting in the bottom of the dustbin. The edges of the dustbin were smooth, so they couldn't get a grip to climb out.

At last, I had a captive audience. I stood over the dustbin and started to talk to the rats. I was firm and said I had had enough of their destructive ways. I explained that I had tried everything to sort this problem out politely, but now I had to be far more strict and severe with my communication.

I told them that listening to me was for their own good as the cats and dogs might kill them if they remained in the feed room. For this reason, they needed to find somewhere else to live. I told them that I would take them to the bottom of the property at least one hundred metres away. I heard that to keep rats away, they need to be moved at least fifty metres away from where they are to keep them from wandering back.

I spoke with the rats for twenty minutes before I called Thomas, one of my gardeners, and asked him to please take the rats down into the field and release them. I told the rats that there was lots of food in the field, and they would not go hungry and could make a pleasant and happy life for themselves there.

Thomas picked up the dustbin, put the lid back on and carried it down to the bottom of the property, where he placed it on its side and opened the lid, letting all the rats run free. He watched all the rats disappear into the long grass.

I quickly said a few words of thanks to the rats for listening to me and went back into the house.

For the next two years, we had no rat issues at all. It seems that every two years, I might need to have a rat conference and repeat this exercise, but that is a small price to pay to keep the rats out of the horse feed.

CHAPTER 7
The True Connection

All animals are sensitive to spirit. They recognize subtle energies at play within their environment, and by watching them, we can become more alert to the presence of spirit around us.

Most traditions teach that domestic animals are not true power animals. They recognise that domestic animals could have significance but that the true totem is their wild counterpart. Humans worked with animals until they could handle the more primal energies in order to domesticate these animals. Therefore, certain traditions can only give totem status to wild animals. For instance, the dog would not be the true totem. You would look to the wild member of the canine family, such as the wolf, African wild dog, etc. For the cat, you would look to the tiger, leopard, lion and so forth.

It is not to say we don't learn from our domestic animals; there is a lot to learn from everything we experience in life.

Shamanism teaches us that we have lessons to learn from all life forms. This is especially true of the animals that are closest to us. All animals, wild and domestic, have unique qualities and personalities, so when we honour our animals, we also celebrate the primal essence behind them.

To transition from communicating with domestic animals to wild animals, we need to examine how we relate to and treat each differently; only then can we begin to relate to them more similarly and effectively.

We personalise our animals by giving them names and using them as our sounding boards, often talking to them as if we were talking to a friend. We feel that they understand us, which they do. However, because most people don't get the chance to converse with wild animals in this way, we feel that they can't possibly understand us. If we learn to talk to wild animals in the same way we speak to our domestic animals, we would be amazed. Wild animals will no longer be mere spectacles, but we will see them as spiritual creatures of great power and complexity.

It does not mean that if you have a great rapport with your domestic animals that the wild kingdom will automatically open up to you. Nor does it mean that by giving them a name, they will open up their power and become pet-like. However, we will start to notice the communications that come to us daily from the less domestic side of nature. We will begin to understand wild animal encounters more fully when we understand our domestic animal behaviour and start to apply that to wild animal behaviour and then see similarities. This returns us to our instinctive ties to the rhythms and lives of the natural world.

In the wild, the leader of the wolf pack goes on a hunt and comes back, sometimes days later, with its prey. The pups run up to see what the leader has for them. This instinctive behaviour also happens with the domestic dog. Domestic dogs get excited on our return from shopping and will often jump up to see if we brought anything from our "hunt".

It is a good idea to always bring our canine friends a small gift after a long trip away. When we go away, in the minds of our animal companions, we are going on a hunt. If we come home empty-handed, they think we are useless as we have been on a hunt and obviously caught nothing. Maybe we will have better luck next time, they think.

Our domestic animal friends have become so integrated into our world that our energy has contaminated them. I use the word "contaminated" because our energies are interlinked, so when we are ill or stressed, our animals pick this up and also become stressed or sick. They end up taking on our issues, and this is what has contaminated them.

On the other hand, the wild animal does not live in a domestic situation and therefore does not have the same relationship with humans. They are still functioning in the state of "raw energy", energy that has not been affected by human contact. For this reason, wild animals are highly connected to nature and the energy of the planet.

Communicating with a domestic animal of any species, while still being spiritual, will give you information of a practical kind. The animal will share with you their likes, dislikes, and information on the stress they pick up in the home and with everyday things. The wild animal, though, will share spiritual information and tell you things going on around the planet.

Some people ask, "What do animals think about all day? Only food? Are they intelligent? Do they remember the past? Can they reason?"

In humans, these are questions usually asked of different races, other cultures, or groups considered different, inferior, or less intelligent. How do people who ask these questions judge animal intelligence? They usually expect animals to prove their intelligence by using the same language, expressions, or symbols that we do; however, animals don't see, think, act, or perceive in the same way. Animal intelligence or ability must be perceived and understood in its own context and on its own terms.

After having communicated with thousands of different species of animals, I find it fascinating that some people still don't see animals as intelligent and aware individuals.

As examples of how we perceive the animal kingdom, a saying like "Bird Brain" gives us the impression that chickens are stupid. If a rabbit runs in front of a car, it must be stupid; doesn't it know it will be run over? Cows are seen as grazing machines. All they do is eat; they don't think, just eat. We are judging them on what we assume is intelligent behaviour.

Try putting yourself in their fur coats or feathered bodies for a moment. We have dulled our senses to such an extent that we are almost immune to the sound of traffic or the brightness of headlights approaching. Animals, however, have an extraordinarily keen sense of vibration, sound, and sight. The vibration of the traffic is deafening to any animal. The headlights of a motor vehicle are startling and scary. The sound and sight of an approaching car puts them in a state of panic, sometimes causing them to freeze or dart in front of the approaching vehicle.

Is this behaviour intelligent? If you lie on the ground and see the world from an animal's perspective and tune in to their hearing, you will feel the fear they feel, and you might react the same way. It's got nothing to do with intelligence – it has to do with animal instinct.

CHAPTER 8
Communicating with the Little Ones

I want to focus a little on the animals or insects we call pests because it is necessary for us to expand our knowledge and understanding of all living things. We need to be much more aware and open our eyes to every aspect of communication.

Accepting and seeing every insect, animal or plant as a spiritual being will help you brush away the edges and be far more open and successful in communicating with them. Even the smallest insect has an intelligence of its own. Our aim here is to reach a far more profound understanding of how they see and experience the world.

As I mentioned previously, the first step to understanding is learning about their habits and how they communicate with their own species. Once you have understood how they relate on a physical level, you will then have a much deeper understanding on a telepathic level. If you don't have a basic idea of insect behaviour, you will not be able to connect effectively on a telepathic level.

For this section, I have singled out a few different small species.

Let's start with the **Snail.**

We see snails as pests that eat all our plants and destroy our vegetable patches, but snails are extremely gentle and sensitive creatures. They hear or feel sound vibrations through their entire bodies. Every pore in their body is highly sensitised to give them a "feeling" picture of the world.

They are the ultimate teachers of experiencing life to its fullest. When they eat, they become totally absorbed in what they are doing. They become one with what they are eating, whether a juicy lettuce leaf or a pile of old cabbage they find in the garbage. They are also highly sexed beings and experience this to a far greater extent than any other animal or human.

Snails don't seem to see with the same kind of visual receptors we do. They don't see solid forms; instead, they sense energy waves so that other animals are shaped according to their body form, including the energy emanating from their bodies. We appear to them not as solid bodies but rather as bands of heat or coloured patterns with smooth or sharp energy projections created by our movements or thoughts.

We expand when we embrace other creatures' ways of sensing and thinking.

The **Rat** is also a creature that is generally misunderstood. They are highly intelligent and relatively clean animals. Rats as pets are usually friendly. They are thought of as architects, constantly rebuilding and refurbishing their homes. Their front teeth never stop growing, and for this reason, they chew everything – including concrete – in an effort to keep them filed down. When you connect with a rat, you will become aware of their extreme intelligence, awareness, sensitivity, and energy.

Mosquitoes don't receive messages in the same way other animals do. All our thought forms come across to them as vibrations, so they pick up the vibration of our thoughts. Mark J. Klowden, a mosquito behaviourist at the University of Idaho, says that "The mosquito has antennae on the top of their heads with sensory receptors. The vibration is picked up via these sensory receptors, sent to the brain which interprets the sensation as sound."

It needs someone with highly developed intuition to first hear their messages and, secondly, interpret them accurately. The intensely annoying high-pitched buzzing of a mosquito in our ears is the most beautiful song to another mosquito. The female uses this sound to attract the male.

Caterpillars are immature butterflies. They signal their presence by vibration, which is caused by rubbing their special organs called vibratory papillae. These signals call the ants, which then gather sugar that caterpillars produce by rubbing the glands on their backs. These glands also produce an odour to alert the ants to predators. Ants act as bodyguards to caterpillars, swarming around them to protect them from wasps and other hunters. Sometimes the ants carry the caterpillar into their nest and feed it there.

Since ants use vibration to call other ants, caterpillars use vibration to call ants. If they are in a tunnel that has caved in, they will call the ants, who will then come to their rescue and dig them out.

Let's look at the **Fly**. If insects were the members of a rock band, the fly would be the drummer. Their communication takes the form of a steady drum rhythm. In this way, they also respond to the vibration of a steady beat. Flies are sociable insects. If you connect to a fly and ask him why he likes to irritate you so much, he will probably tell you that he is not trying to annoy you but rather be your friend.

Yes, you can communicate with flies. They do hear the vibration of your thoughts. However, it is not as easy for humans to hear and interpret their thoughts. Communicating with insects takes a lot of concentration and practice. Developing your intuition to the ultimate level is an excellent place to start.

So, what of the **Spider**. The spider helps you develop and weave your path in life. It shows you that you need to focus on your creativity and sensitivity.

Spiders show the ultimate state of patience and beauty. A spider in your home reflects the changes you need to make in your life to proceed with confidence without getting caught up in what others think of you and what you are doing.

Spiders are incredibly sensitive and show us how we should weave our own intricate webs of beauty in our lives while allowing our dark side to be hidden. Often a spider in the home will reflect on what is going on inside our own minds; our fears, insecurities and dark thoughts are all brought out by the spider, ultimately leaving us with the joy of life and appreciation of those around us.

The spider's intricate web of beauty is something we can all strive towards in developing the beauty in our lives.

The **Ant,** for its size, is the most powerful being on the planet. It can carry things three times its size without any problem. Ants are one of the busiest animals around and never sit still. Always looking out for each other, they show us working as a team is a good thing.

When communicating telepathically with ants, it is always better to focus on the queen ant because the other ants will follow whatever she says.

Ants are incredibly family conscious and do everything together. We can learn a lot from them about connecting with family and friends and how to work as a team.

Bees buzzing around you is a sign of good luck and the prospect of abundance coming into your life. They teach us to look at the beauty of the world around us and help us to find the sweetness in our world and our life. They are incredibly gentle and totally focused on their purpose.

A few years ago, we had a massive swarm of bees move into a spot under a Wendy House next to our home. After having them there for many months, I began to worry about the dogs and horses getting stung by the bees, so I organised a beekeeper to come and fetch them. I made him promise not to kill or injure any of the bees.

He arrived a few days later and, together with his helper, managed to safely collect all the bees and take them to his farm, where he would keep them safe. He suggested we put a bee box high up in the tree so that when another hive decided to move in, they could move into the bee box, making it much easier for him to fetch them the next time.

A few months later, the box was full of bees, so he came out to fetch it and replace it with another box for the next lot of bees that arrived. I should mention that we live in a "bee belt", so we often get swarms of bees passing through.

One day, I was sitting under the tree where the bee box is housed. A bee found me and decided to come and see what I was doing. She buzzed close to my face and wouldn't move away. I told her I felt uncomfortable with her buzzing so close and asked her to please move away, but she still stayed there.

Suddenly I felt this calm and relaxed feeling come over me, and I asked her if this was coming from her. "Yes," she said, "I am feeling calm and relaxed around you."

All I could do was tell her that I loved and respected her and how beautiful I thought she was. With that, she suddenly moved away and went to join the rest of her hive.

A **Butterfly** symbolises beauty, change, resurrection, strength, growth, and courage.

A butterfly has a deep sense of belonging and understanding that nothing is impossible. It radiates beauty and shows you that change is essential in life. Without change, you can't move forward, and you will become stagnant.

By moving forward with confidence and strength, your own beauty will shine through, you will grow, and life will have far more meaning to you and those around you – this is the message from the butterfly.

CHAPTER 9
The Human-Animal Bond

Over the centuries, there have been many stories of the bond between humans and animals. We have come a long way in becoming more aware of the special abilities and importance of the animals in our lives.

In 1982, a centre was established at Purdue University, later renamed the Centre for Human-Animal Bond. They studied the relationship between people and the animals they shared their homes with.

During their studies, they found that the relationship between humans and their animal companions was so dynamic that it had profound effects on the psychological and physiological condition of each other. It was found that people living with animals generally had a marked decrease in blood pressure, lower anxiety levels and were far happier and healthier. Children who observed the behaviour of animals often learnt to become more nurturing and made better parents.

Over the years, I have studied the effects animals have on people. I have noticed how animals have an incredible impact on increasing the confidence and self-image of both adults and children. You just need to take note of the effect animals have on the people around them when they are taken into a hospital, retirement village or orphanage. Today many groups run Animal-Assisted Therapies with wonderful results.

I have also found a definite link between domestic violence and animal abuse. It starts with a lack of self-esteem, resulting in anger which is taken out on the people and animals closest to the perpetrator.

There is always an emotional or physical reason behind an aggressive animal, as the animal is not born that way. It will become aggressive if it is insecure, scared, or in pain. The same applies to a human who feels inadequate, insecure, suffering from a lack of self-esteem or being in pain, either physically or emotionally. They, too, will become aggressive and sometimes even abusive.

One of the many reasons we are drawn to animals more than humans is that animals are genuinely unconditional. Their love is unconditional, they have no expectations, they don't judge how we look or behave, and they are always happy to see us. Have you ever met a human who can live by these standards?

I am often asked what the most aggressive species of animal I have ever worked with is. My answer is, without a doubt, the "Human", who is by far the most aggressive species I have ever encountered.

When a human has a negative reaction towards another person or situation, more often than not, the animal living with this human will have the same reaction. The animal is literally reflecting the response of the human. If the human dislikes a specific person, the animal will also build up similar feelings.

Animals make the best judgement of people, so if an animal doesn't like a certain person, there is always a good reason; you need to be aware and take the proper precautions where this person is involved.

PURR-FECT VOICES

ANIMAL'S PRAYER OF LOVE

When you are calm, and you are still
Know that we are waiting till
the sun comes up, and the day begins

We ask you to guide us through our day
And help us find the time to play.
Do not stress and do not weep
Enjoy your time
Enjoy your sleep
We might be small
We might be tall
We might have fins or fangs or fur
Some of us can even purr

Protect us as we live our lives
and help our human friends get wise
Let them see the heart inside
And feel the love we give with pride

We ask you to guide us through our day
And touch us in a loving way
We are here to help you through your day
And teach you how to laugh and play

JENNY SHONE

Another chance to rise and sing
To show our love from deep within
A love that surpasses everything

— CHANNELLED BY JENNY SHONE.
DRAWING BY LINDA BROWN

CHAPTER 10
Developing your Intuition

To reach the ultimate level of communication, it is vital to develop your intuition.

In the centre of the forehead lies the pituitary gland, behind which is the pineal gland. The pineal gland lies dormant most of our lives. Once you clear the toxins and blockages away from the pituitary gland and start focusing on the pineal gland, your intuition will develop to a much deeper level.

To access your intuition, you need to find your still point. Finding your still point means reaching a point of total vulnerability and becoming neutral. It takes a lot of practice. The easiest way to access your still point is to focus on your breathing until you feel as if you are floating.

Watching the news on television or listening to it on the radio will fill your mind with negative energy. However, if you can hear or see the chaos in the world around you, acknowledge it and let it go – you will then be able to tap into your intuition successfully.

If you let all the negative news affect you and fill your mind with worry, it will fill you up like a jug filling with a thick dark liquid leaving no space for intuition and preventing you from successfully reaching more profound levels of communication.

It is essential to clear and stimulate both the pituitary and pineal glands and start with as clean a slate as possible. Healthy eating, an exercise plan, and regular meditation focusing on the above two centres will be of utmost importance and get you to the point where you can successfully access your telepathic abilities.

STIMULATING THE PITUITARY & PINEAL GLANDS

To stimulate your pituitary and pineal glands, it is a good idea to start by doing a relaxation exercise starting from the top of your head, slowly moving down your body to the tips of your toes. Focus on your breathing and allow yourself to let go totally. Release any stress, tension, or concerns. Feel yourself floating on air, becoming completely light.

Visualise a bright light the size of a marble in the centre of your forehead. Watch this marble of light get brighter and brighter and at the same time feel it get warmer and warmer. See the light getting clearer and clearer. When this is a clear, bright light, start seeing another marble of light just behind but slightly above. Watch this light getting brighter and brighter as it gets warmer and warmer.

Keep focusing on these marbles of bright light for a while. Now be aware of the two shining marbles in the centre of your forehead; focus on these two lights. These marbles of light are not stagnant; they are fluid marbles of energy swirling around in the centre of your forehead. They are alive – feel their energy as you sit for a few moments.

Slowly start feeling the skin on your face, feel the breath going in and out of your lungs. Feel and hear the sound of your heart beating. Move your fingers and toes before opening your eyes.

RELAXATION MEDITATION

Take fifteen minutes off daily, and put a sign on your door saying, "Do not disturb – Busy relaxing." Set a timer for fifteen minutes, sit quietly, and focus on your breathing. Become aware of only your breathing and nothing else. Enter into a space of nothingness and just relax. When the timer rings, start to feel your body and slowly return to your normal state.

When developing your intuitive abilities, it is important to realise that some people are clairvoyant, others might be clairsentient, while others are clairaudient. The first thing you need to do is establish which one you are.

If you discover that you are clairsentient, then focus on being that and developing your strong suit. It is far better to develop what you are already good at rather than focusing on the other psychic senses and becoming a jack of all trades and master of none.

Once you realise your strength, the other senses will come.

Focus on what comes naturally to you. Not everyone is the same, nor is every animal the same. Whichever intuitive power comes naturally to you, work on strengthening that trait. Allow your natural ability to be your strength.

The tenser you are, the more you will get in your way. Relax, let yourself go, and enjoy tapping into your natural abilities.

Develop your Clairvoyance by creating a holographic image

Find your still point by doing a two-minute relaxation focusing on your breath and then accessing your imagination.

Start by picturing a juicy red apple in your mind. When you can see it, try to feel the texture of the fruit. Picture yourself taking a bite of the apple and tasting its sweet juice. Smell this crispy, sweet red apple.

Now picture the apple being on the chair next to you. Let it become real. Once you can see it on the chair, imagine it on top of a table on the other side of the room. Move the apple around the room. When you are finished, slowly let the apple disappear and return to your normal state.

This exercise can also be practiced with your dog. Wait until the dog is sleeping, then picture one of its favourite treats. Run through the same process as the apple. When you are ready, picture the treat on the ground right in front of your dog's nose. See how long it takes for him or her to wake up and search for the treat. Remember, once the dog wakes up, always give them the same treat you just pictured.

Develop your Clairaudience by letting your imagination work for you

When your animals come up to you, imagine they are talking to you. What kind of voice do they have? How do they sound? Do they have a high-pitched or gruff voice?

Sit in the garden and relax. Now imagine the birds are talking to you. What do they sound like? What are they saying?

Practice with the neighbour's animals. How do they sound? How different is the cat's voice from the dog's?

The more you activate and use your imagination, the easier it will be to hear the voices of the animals – remember to make it fun.

Develop your Clairsentience by tuning in to the feelings and energies of the animals around you

Be aware of the feelings and sensations in your body before you start this exercise so you can distinguish between your emotions and that of the animal you are practicing with.

Sit in your vet's consulting room or even in a dog park and when a dog comes past, close your eyes and feel the dog's emotions before opening your eyes to assess the situation.

Practice feeling the emotions of any animal that approaches you without looking at their body language. Feel what emotions lie behind the obvious behaviour of the animal.

When an animal approaches you, don't think, instead feel what is going on in their minds. Do you feel anger, insecurity, happiness, or laziness? Do you feel the animal is sick and depressed? What else can you pick up from this animal?

An excellent place to practice this exercise is at a stable yard. There are lots of horses who are not known to you. Feel the emotions in the yard. When you have practiced with a few horses, get some feedback by asking someone why a certain horse is feeling angry? Or why another horse might be lonely?

The more positive affirmation you receive, the more confidence you will get and with confidence comes strength. Your intuitive abilities will grow. You will become more and more accurate with your readings.

Remember – the more you put in, the more you will get out.

CHAPTER 11
Helping Missing Animals Find Their Way Home

Connecting and working with missing animals is by far the most difficult and stressful part of being an animal communicator. When connecting with a missing animal, you are dealing with not only the stress you feel coming from that particular animal, but you are also dealing with the stress you feel coming from the human companion.

It is important to note that most animals who have gone missing are not lost. They have a good idea of where they are due to their in-built GPS. They can see the earth's energy, which is how they map their journey back home. Animals also use the scent that they leave to help navigate. However, if an animal is subjected to severe stress, is injured, abused, or stolen, this can disturb them, making it difficult, sometimes impossible, to find their way home.

When working with a missing animal, it is always necessary to first determine if the animal is injured or has crossed over. For an animal communicator to establish if an animal has crossed over is not easy. Once they have crossed, if you ask them how they are, the animal will always respond that they are alive and well because, in their soul form, they are very much alive. Sometimes they don't even realise that they have crossed. While connected to them, you might feel some of the tension, fear, or anxiety they felt just before crossing over.

Over the years, we have developed some techniques to be able to discover if they are still here with us in the physical dimension or are on the other side in a different space.

In the following few stories, Sandy Whitfield Kurn shares her experiences during her many years of working with missing animals, and you will hear her speak of placing an animal in protection. This she does by visualising the missing animal surrounded by a white bubble of light. The bubble forms a shield around the animal, and often in this state, the animal will guide themselves home, provided they have not been involved in an accident or being held captive somewhere else.

I trained Sandy, and she worked closely with me for many years, helping me run my workshops all around the country. With my guidance, she perfected the art of working with missing animals and today handles all my missing animal cases for me.

The following four success stories are told by Sandy.

THREE MISSING DACHSHUNDS

Liza was panicking over her three little Dachshunds. She and the family had recently moved to a new house, and the dogs had escaped from the garden. Liza was at work when a neighbour informed her that the dogs had got out and were now missing. Liza called me, and I immediately put the white bubble of protection in place and looked for one of the dogs to connect to. Fido, the male, was the pack leader, while the other two were girls.

The three little dogs had gone for a run. Not knowing the area, they had gone into a mealie field behind the new house.

We managed to guide Zoey back home first as she had left the other two and wandered off. She arrived home in the early hours of the following day. When I connected with Fido, he was frightened as he couldn't see anything beyond the front of his nose. All he could hear was the noise of the wind blowing the mealie plants.

The protective bubble was in place on Fido and Xenia. In my mind, I pictured a silver cord attached to them so that I could help guide them to the road. Liza and her sister went by car to the edge of the mealie field and drove around looking for the dogs. The car was running low on petrol, so they had to rush off, but when Liza returned, out of the mealies came two little dogs, heading straight to Liza's car.

Happy Mom and three tired dogs.

GETTING BELLA HOME

It was late one Sunday afternoon when I received a call from Becky. Her Jack Russell had run off during the morning into a nearby field, and Becky had been driving around calling her dog for close to four hours when she phoned me.

I connected with the little dog named Bella. She was still lost in the overgrown field near her house, and the only sounds she could hear were the wind and the plants rustling around her. She couldn't hear Becky. She was small, with the plants so tall, and it was quite dark down on the ground. The poor little girl whimpered in frustration as she didn't know which way to go.

I asked Becky to park the van at the edge of the field and not to move but wait and watch for Bella to come to her. I put Bella in a bubble of white light for protection and attached a bright white lead to the bubble. I then asked Bella to follow wherever the lead pulled her to get her back to Mommy. Shortly after that, Bella scampered out of the field, straight to Becky.

We had a happy Bella and happy Becky. It was a wonderful ending to a frightening few hours.

A MISSING KITTEN

On a Saturday morning, a distraught Leigh-Anne called to say that one of her little black kittens had gone missing. As previously mentioned, when we hear of an animal gone missing, the first thing is to put the bubble of protection around them. This I did with little Lillea.

Leigh-Anne and her husband have a private rescue sanctuary housing about thirty cats. They treasured them all. However, Leigh-Anne quickly noted that one was missing, so I was called in to help get the kitten home safely.

Having put protection around Lillea, I introduced myself to her and reassured her that I was there to help her find her way home. I attached a long silver cord to show her the way back. Lillea was a star, and we got her home early in the evening that same day. She was none the worse for her walkabout but pleased to be with her siblings and Mommy.

BENJAMIN'S TALE

Benjamin is a lovely gentleman of a cat. Although he was a homely boy, he had had enough of walking sedately with Mom and Dad and decided to go on an "adventure". Benjie was on holiday with his humans, Ophelia, and Howard, in a small town in the Cape when he went off on his own and didn't come back when called. Out came the bubble of protection, and I started talking to Benjie.

He told me he was having a wonderful time and intended to go home to Mom and Dad, but only when he was ready. I asked Benjie to show me where he was, and all he could show me was a long patch of grassy veld. He loved that the wind was blowing and said it reminded him of days past when he was free. Benjie, it seemed, came from a feral life to live with his Mom and Dad in town.

The end of the holiday was looming, and Benjamin was still missing. We were all getting desperate.

The next time I connected with Benjie, he showed me that he was on a hillside above the town. The main thing we could see was a church; the rest was houses and little roads. I determined that Benjie could understand what a church was, and I suggested he went there when the sun was setting where Mom and Dad would be with food.

On the first night, Benjie didn't return. The night before Howard and Ophelia were due to go home to Cape Town, which was too far to come back in a hurry, Benjie arrived. To everyone's relief, he had recognised the area and had managed to meet up with Mom and Dad. Everyone was extremely emotional and thrilled to have their little boy back.

STOLEN HUSKY

This is an experience I had while assisting Catherine, one of the ladies who helped me locate missing animals. I had trained Catherine, and one day she came across a case she needed help with.

It was late one day when I received a call from Catherine, and she asked me if a dog could change the colour of its eyes while communicating on a telepathic level.

Catherine had been asked to try and locate a husky that had been stolen in the Cape. The photograph of the husky showed she had blue eyes, but when Catharine connected, the husky who came to her had brown eyes.

I told her it was impossible for a dog to change their eye colour during a communication session and offered to help her see if I could find out what had happened to this husky.

What I discovered was that she had been connecting with a different husky to the one she was meant to. When I connected, I saw there were a few Huskies all together in a small room. On deeper investigation, I found there was a syndicate who had been stealing Huskies and keeping them in a small, closed, stuffy room to sell to anyone who would buy them.

While I was connected, I could hear the sound of aeroplanes, so I realised these Huskies were being kept near the Cape Town airport. I gave all the information to the lady whose dog was missing and asked her to give it to the police to follow up and investigate the situation. In the meantime, I placed all the dogs in bubbles of protection.

The police informed her that they had suspected a syndicate operating in that area. With the information from their ongoing investigation and the little information we gave them, the police managed to find the place near the airport and rescue all the dogs. This was indeed a happy ending to a potentially disastrous situation.

THE IMPORTANCE OF THE WHITE LIGHT

A friend of mine decided to go away for the long weekend. He had three cats, all just under a year old, and decided to place them all together in the kennels to keep them safe while he was away.

While cleaning their little garden, a cleaner at the kennel accidentally left the gate open. Inevitably one of the little cats got out. When the staff came to feed, they noticed one of these cats missing and frantically looked for her.

I was called to help get this kitty back safely. Since I was quite far away, I could only work from a distance. The first thing I did was picture the cat surrounded by a white light to keep her safe. I then connected with her and found she had run off into a field of burnt grass next to the kennels and was hiding there. She was very anxious and didn't want to go too far away from her siblings. I explained to her that the kennel was her safe place, and it was vital that she go back to where her brother and sister were waiting for her. I told her there would be food for her, and she would be kept safe and warm until her people came to fetch all of them to take them home in a few days.

Later that evening, the kennel staff heard a soft mewing and went out to find this little kitty sitting next to the fence where her brother and sister were. They managed to catch her and put her back in her garden with the others. All of us were relieved.

GUIDING A YOUNG EAGLE HOME

Sundays are very busy at the Bird of Prey Centre. One of the displays showed how a young eagle would fly off and, when called, would return to sit on his handler's arm, where he would receive food treats. On this particular day, the eagle flew off eagerly when suddenly a wild eagle appeared. This completely derailed the young eagle, who flew off into the distance and didn't come back when called.

It is almost impossible to locate a bird in flight as the trees and the sky all look the same. Even the rooftops look similar, so you cannot pinpoint any landmarks. All I could do was put a picture in my mind of the eagle coming back.

It took about three hours, but eventually, we saw a bird in the distance coming in our direction. As it got closer, we noticed it was the eagle coming home. He had seen the picture of himself coming back and followed this thought coming home all on his own.

It doesn't, however, always happen as successfully as this day. Sometimes animals just choose not to listen, but other times, they listen, and we can all breathe easily again.

CHAPTER 12
The Spiritual Side of Animals

All animals are spiritual beings, and they've been seen that way for centuries. Sometimes they have even been worshipped as gods or seen as spirit guides and animal totems. They teach us how to love and feel loved. They have the most extraordinary way of elevating us and helping us to appreciate nature unconditionally.

Animals are incredibly connected to nature and the planet. However, they sometimes forget that they are spiritual beings, and this usually happens if they are neglected, abused, or abandoned. In these cases, all it takes is for us to see the animal as a spiritual being and point this out to them. They can then start activating their spiritual side and begin to grow spiritually at an alarming rate.

I had first-hand experience with this when I adopted Riff-Raff. He came to me as a healthy, fun-loving, and boisterous three-month-old puppy. He was instrumental in helping me start The Animal Healing Centre and worked side by side with me for many years. I would regularly meditate and work telepathically with animals, and as a result, Riff-Raff found his own spirituality and developed exponentially. Within a few months, he had grown so much spiritually that he overtook me on his spiritual journey.

I later discovered that the more you acknowledge an animal as a spiritual being, your telepathic connection will become deeper.

For an animal to be spiritual means that they have a deep understanding of their importance in the universe and then discover their purpose in our lives.

Animals can lead us spiritually in many different ways. They teach us about death, participate in our social and moral development, enhance our physical and psychological well-being, and heighten our capacity to love and experience joy. To allow for this spiritual growth, we need to open our hearts and allow love to flow in and out so that everyone around us, including the animals, can feel our love. It is important to remember that all our interactions with animals need to be done with ultimate respect, love, compassion and understanding.

One of the ways to see how animals can be spiritual beings is to recognise how they help people with mental, emotional, and even physical issues. They uplift and bring joy to everyone that will allow them to. They have a profound understanding of life and even death and are often depicted in visions of the afterlife, seen as guardians and thought of as the heart and soul of the universe.

We can benefit spiritually in our relationship with animals; they give us the fundamental and immediate sense of both joy and the wonder of creation. Animals seem to feel more intensely and purely than we do because they have such a strong and healthy connection to everything around them.

Animals teach us how to love, how to be loved and how to enjoy life to the fullest. They teach us how to feel. They teach us how to discover our inner feelings and help us find our purpose in life. They teach us the language of the spirit. They communicate with us in a language that does not require words. They help us understand that words might even stand in the way of true communication.

So the next time you look at a dog, cat, horse, or even wild animal, look beyond the physical and see the soul inside the animal, then acknowledge them in a deep and understanding way and just watch how they grow and develop spiritually right in front of your own eyes.

About the Author

Since childhood, Jenny has had the ability to talk to animals on a telepathic level. She grew up in a family that loves animals and has always supported her in developing this ability.

Jenny is an active individual. She is a keen horse rider and has worked with the world-famous Lipizzaner horses in Kyalami; she also visited Vienna's Spanish riding school. Jenny has always loved music and dancing. While living in Natal, she took up modern dancing and later qualified as an international aerobics instructress, where she took part in the 9-hour aerobics marathon and travelled to various aerobic seminars around the country.

Jenny has also served her community by joining the police reservists, where she helped fight crime and protect animals in need. This she did for over fifteen years and eventually resigned as a police inspector to fulfil her dreams of animal communication and healing.

She founded The Animal Healing Centre in 2002, where she specialised in all types of energy healing, including telepathic communication with all species of animals. Jenny and her business partner travelled all over the country, presenting workshops in animal telepathic communication and introducing animal communication to Zimbabwe and Namibia. She was also instrumental in running fundraising events for various animal charities.

Jenny was nominated for Woman of the Year in 2007 and was one of twelve professional animal communicators selected around the world to speak at the World Animal Speak Summit in 2014.

She spends a lot of her time consulting with animals worldwide, helping their people develop a much deeper understanding of what the animals want and need to help keep them fit, healthy and happy.

Her favourite place to visit is the Kruger National Park, where she enjoys spending time with the animals in their wild habitat.

She shares a home with her husband, four dogs, a cat, four donkeys, a miniature mule, and two horses.

It is Jenny's calling to improve the lives of all the animals that cross her path.

www.ingramcontent.com/pod-product-compliance
Lightning Source LLC
Chambersburg PA
CBHW072011290426
44109CB00018B/2207